CW00458117

Letters on Astr

In which the Elements of the Science are Familiarly
Explained in Connection with Biographical Sketches of the
Most Eminent Astronomers

Denison Olmsted

Alpha Editions

This edition published in 2022

ISBN : 9789356783195

Design and Setting By
Alpha Editions
www.alphaedis.com
Email - info@alphaedis.com

Contents

LETTER I.

INTRODUCTORY OBSERVATIONS.

"Ye sacred Muses, with whose beauty fired,
My soul is ravished, and my brain inspired,
Whose priest I am, whose holy fillets wear;
Would you your poet's first petition hear,
Give me the ways of wandering stars to know,
The depths of heaven above, and earth below;
Teach me the various labors of the moon,
And whence proceed th' eclipses of the sun;
Why flowing tides prevail upon the main,
And in what dark recess they shrink again;
What shakes the solid earth, what cause delays
The Summer nights, and shortens Winter days."

Dryden's Virgil

To Mrs. C——M——.

Dear Madam,—In the conversation we recently held on the study of Astronomy, you expressed a strong desire to become better acquainted with this noble science, but said you had always been repelled by the air of severity which it exhibits, arrayed as it is in so many technical terms, and such abstruse mathematical processes: or, if you had taken up some smaller treatise, with the hope of avoiding these perplexities, you had always found it so meager and superficial, as to afford you very little satisfaction. You asked, if a work might not be prepared, which would convey to the general reader some clear and adequate knowledge of the great discoveries in astronomy, and yet require for its perusal no greater preparation, than may be presumed of every well-educated English scholar of either sex.

You were pleased to add the request, that I would write such a work,—a work which should combine, with a luminous exposition of the leading truths of the science, some account of the interesting historical facts with which it is said the records of astronomical discovery abound. Having, moreover, heard much of the grand discoveries which, within the last fifty years, have been made among the *fixed stars*, you expressed a strong desire to learn more respecting these sublime researches. Finally, you desired to see the argument for the existence and natural attributes of the Deity, as furnished by astronomy, more fully and clearly exhibited, than is done in any work which you have hitherto perused. In the preparation of the proposed treatise, you urged me to supply, either in the text or in notes, every *elementary principle* which would be essential to a perfect understanding of the work; for although, while at school, you had paid some attention to geometry and natural philosophy, yet so much time had since elapsed, that your memory required to be refreshed on the most simple principles of these elementary studies, and you preferred that I should consider you as altogether unacquainted with them.

Although, to satisfy a mind, so cultivated and inquisitive as yours, may require a greater variety of powers and attainments than I possess, yet, as you were pleased to urge me to the trial, I have resolved to make the attempt, and will see how far I may be able to lead you into the interior of this beautiful temple, without obliging you to force your way through the "jargon of the schools."

Astronomy, however, is a very difficult or a comparatively easy study, according to the view we take of it. The investigation of the great laws which govern the motions of the heavenly bodies has commanded the highest efforts of the human mind; but profound truths, which it required the mightiest efforts of the intellect to disclose, are often, when once discovered, simple in their complexion, and may be expressed in very simple terms. Thus, the creation of that element, on whose mysterious agency depend all the forms of beauty and loveliness, is enunciated in these few monosyllables, "And God said, let there be light, and there was light;" and the doctrine of universal gravitation, which is the key that unlocks the mysteries of the universe, is simply this,—that every portion of matter in the universe tends towards every other. The three great laws of motion, also, are, when stated, so plain, that they seem hardly to assert any thing but what we knew before. That all bodies, if at rest, will continue so, as is declared by the first law of motion, until some force moves them; or, if in motion, will continue so, until some force stops them, appears so much a matter of course, that we can at first hardly see any good reason why it should be dignified with the title of the first great law of motion; and yet it contains a truth which it required profound sagacity to discover and expound.

It is, therefore, a pleasing consideration to those who have not either the leisure of the ability to follow the astronomer through the intricate and laborious processes, which conducted him to his great discoveries, that they may fully avail themselves of the *results* of this vast toil, and easily understand truths which it required ages of the severest labor to unfold. The descriptive parts of astronomy, or what may be called the natural history of the heavens, is still more easily understood than the laws of the celestial motions. The revelations of the telescope, and the wonders it has disclosed in the sun, in the moon, in the planets, and especially in the fixed stars, are facts not difficult to be understood, although they may affect the mind with astonishment.

The great practical purpose of astronomy to the world is, enabling us safely to navigate the ocean. There are indeed many other benefits which it confers on man; but this is the most important. If, however, you ask, what advantages the study of astronomy promises, as a branch of education, I answer, that few subjects promise to the mind so much profit and entertainment. It is agreed by writers on the human mind, that the intellectual powers are enlarged and strengthened by the habitual contemplation of great objects, while they are contracted and weakened by being constantly employed upon little or trifling subjects. The former elevate, the latter depress, the mind, to their own level. Now, every thing in astronomy is great. The magnitudes, distances, and motions, of the heavenly bodies; the amplitude of the firmament itself; and the magnificence of the orbs with which it is lighted, supply exhaustless materials for contemplation, and stimulate the mind to its noblest efforts. The emotion felt by the astronomer is not that sudden excitement or ecstasy, which wears out life, but it is a continued glow of exalted feeling, which gives the sensation of breathing in a purer atmosphere than others enjoy. We should at first imagine, that a study which calls upon its votaries for the severest efforts of the human intellect, which demands the undivided toil of years, and which robs the night of its accustomed hours of repose, would abridge the period of life; but it is a singular fact, that distinguished astronomers, as a class, have been remarkable for longevity. I know not how to account for this fact, unless we suppose that the study of astronomy itself has something inherent in it, which sustains its votaries by a peculiar aliment.

It is the privilege of the student of this department of Nature, that his cabinet is already collected, and is ever before him; and he is exempted from the toil of collecting his materials of study and illustration, by traversing land and sea, or by penetrating into the depths of the earth. Nor are they in their nature frail and perishable. No sooner is the veil of clouds removed, that occasionally conceals the firmament by night, than his specimens are displayed to view, bright and changeless. The renewed pleasure which he feels, at every new survey of the constellations, grows into an affection for objects which have so often ministered to his happiness. His imagination aids him in giving them a personification, like that which the ancients gave to the constellations; (as is evident from the names which they have transmitted to us;) and he walks abroad, beneath the evening canopy, with the conscious satisfaction and delight of being in the presence of old friends. This emotion becomes stronger when he wanders far from home. Other objects of his attachment desert him; the face of society changes; the earth presents new features; but the same sun illumines the day, the same moon adorns the night, and the same bright stars still attend him.

When, moreover, the student of the heavens can command the aid of telescopes, of higher and higher powers, new acquaintances are made every evening. The sight of each new member of the starry train, that the telescope successively reveals to him, inspires a peculiar emotion of pleasure; and he at length finds himself, whenever he sweeps his telescope over the firmament, greeted by smiles, unperceived and unknown to his fellow-mortals. The same personification is given to these objects as to the constellations, and he seems to himself, at times, when he has penetrated into the remotest depths of ether, to enjoy the high prerogative of holding converse with the celestials.

It is no small encouragement, to one who wishes to acquire a knowledge of the heavens, that the subject is embarrassed with far less that is technical than most other branches of natural history. Having first learned a few definitions, and the principal circles into which, for convenience, the sphere is divided, and receiving the great laws of astronomy on the authority of the eminent persons who have investigated them, you will find few hard terms, or technical distinctions, to repel or perplex you; and you will, I hope, find that nothing but an intelligent mind and fixed attention are requisite for perusing the Letters which I propose to address to you. I shall indeed be greatly disappointed, if the perusal does not inspire you with some portion of that pleasure, which I have described as enjoyed by the astronomer himself.

The dignity of the study of the heavenly bodies, and its suitableness to the most refined and cultivated mind, has been recognised in all ages. Virgil celebrates it in the beautiful strains with which I have headed this Letter, and similar sentiments have ever been cherished by the greatest minds.

As, in the course of these Letters, I propose to trace an outline of the history of astronomy, from the earliest ages to the present time, you may think this the most suitable place for introducing it; but the successive discoveries in the science cannot be fully understood and appreciated, until after an acquaintance has been formed with the science itself. We must therefore reserve the details of this subject for a future opportunity; but it may be stated, here, that astronomy was cultivated the earliest of all the sciences; that great attention was paid to it by several very ancient nations, as the Egyptians and Chaldeans, and the people of India and China, before it took its rise in Greece. More than six hundred years before the Christian era, however, it began to be studied in this latter country. Thales and Pythagoras were particularly distinguished for their devotion to this science; and the celebrated school of Alexandria, in Egypt, which took its rise about three hundred years before the Christian era, and flourished for several hundred years, numbered among its disciples a succession of eminent astronomers, among whom were Hipparchus, Eratosthenes, and Ptolemy. The last of these composed a great work on astronomy, called the 'Almagest,' in which is transmitted to us an account of all that was known of the science by the Alexandrian school. The 'Almagest' was the principal text-book in astronomy, for many centuries afterwards, and comparatively few improvements were made until the age of Copernicus. Copernicus was born at Thorn, in Prussia, in 1473. Previous to his time, the doctrine was held, that the earth is at rest in the centre of the universe, and that the sun, moon, and stars, revolve about it, every day, from east to west; in short, that the *apparent* motions of the heavenly bodies are the same with their *real* motions. But Copernicus expounded what is now known to be the true theory of the celestial motions, in which the sun is placed in the centre of the solar system, and the earth and all the planets are made to revolve around him, from west to east, while the apparent diurnal motion of the heavenly bodies, from east to west, is explained by the revolution of the earth on its axis, in the same time, from west to east; a motion of which we are unconscious, and which we erroneously ascribe to external objects, as we imagine the shore is receding from us, when we are unconscious of the motion of the ship that carries us from it.

Although many of the appearances, presented by the motions of the heavenly bodies, may be explained on the former erroneous hypothesis, yet, like other hypotheses founded in error, it was continually leading its votaries into difficulties, and blinding their minds to the perception of truth. They had advanced nearly as far as it was practicable to go in the wrong road; and the great and sublime discoveries of modern times are owing, in no small degree, to the fact, that, since the days of Copernicus, astronomers have been pursuing the plain and simple path of truth, instead of threading their way through the mazes of error.

Near the close of the sixteenth century, Tycho Brahe, a native of Sweden, but a resident of Denmark, carried astronomical observations (which constitute the basis of all that is valuable in astronomy) to a far greater degree of perfection than had ever been done before. Kepler, a native of Germany, one of the greatest geniuses the world has ever seen, was contemporary with Tycho Brahe, and was associated with him in a part of his labors. Galileo, an Italian astronomer of great eminence, flourished only a little later than Tycho Brahe. He invented the telescope, and, both by his discoveries and reasonings, contributed greatly to establish the true system of the world. Soon after the commencement of the seventeenth century, (1620,) Lord Bacon, a celebrated English philosopher, pointed out the true method of conducting all inquiries into the phenomena of Nature, and introduced the *inductive method of philosophizing*. According to the inductive method, we are to begin our inquiries into the causes of any events by first examining and classifying all the *facts* that relate to it, and, from the comparison of these, to deduce our conclusions.

But the greatest single discovery, that has ever been made in astronomy, was the law of universal gravitation, a discovery made by Sir Isaac Newton, in the latter part of the seventeenth century. The discovery of this law made us acquainted with the hidden forces that move the great machinery of the universe. It furnished the key which unlocks the inner temple of Nature; and from this time we may regard astronomy as fixed on a sure and immovable basis. I shall hereafter endeavor to explain to you the leading principles of universal gravitation, when we come to the proper place for inquiring into the causes of the celestial motions, as exemplified in the motion of the earth around the sun.

LETTER II.

DOCTRINE OF THE SPHERE.

"All are but parts of one stupendous whole,
Whose body Nature is, and God the soul."—*Pope.*

LET us now consider what astronomy is, and into what great divisions it is distributed; and then we will take a cursory view of the doctrine of the sphere. This subject will probably be less interesting to you than many that are to follow; but still, permit me to urge upon you the necessity of studying it with attention, and reflecting upon each definition, until you fully understand it; for, unless you fully and clearly comprehend the circles of the sphere, and the use that is made of them in astronomy, a mist will hang over every subsequent portion of the science. I beg you, therefore, to pause upon every paragraph of this Letter; and if there is any point in the whole which you cannot clearly understand, I would advise you to mark it, and to recur to it repeatedly; and, if you finally cannot obtain a clear idea of it yourself, I would recommend to you to apply for aid to some of your friends, who may be able to assist you.

Astronomy is that science which treats of the heavenly bodies. More particularly, its object is to teach what is known respecting the sun, moon, planets, comets, and fixed stars; and also to explain the methods by which this knowledge is acquired. Astronomy is sometimes divided into descriptive, physical, and practical. Descriptive astronomy respects *facts*; physical astronomy, *causes*; practical astronomy, the *means of investigating the facts*, whether by instruments or by calculation. It is the province of descriptive astronomy to observe, classify, and record, all the phenomena of the heavenly bodies, whether pertaining to those bodies individually, or resulting from their motions and mutual relations. It is the part of physical astronomy to explain the causes of these phenomena, by investigating the general laws on which they depend; especially, by tracing out all the consequences of the law of universal gravitation. Practical astronomy lends its aid to both the other departments.

The definitions of the different lines, points, and circles, which are used in astronomy, and the propositions founded upon them, compose the *doctrine of the sphere*. Before these definitions are given, I must recall to your recollection a few particulars respecting the method of measuring angles. (See Fig. 1, page 18.)

A line drawn from the centre to the circumference of a circle is called a *radius*, as C D, C B, or C K.

Any part of the circumference of a circle is called an *arc*, as A B, or B D.

An angle is measured by an arc included between two radii. Thus, in Fig. 1, the angle contained between the two radii, C A and C B, that is, the angle A C B, is measured by the arc A B. Every circle, it will be recollected, is divided into three hundred and sixty equal parts, called degrees; and any arc, as A B, contains a certain number of degrees, according to its length. Thus, if the arc A B contains forty degrees, then the opposite angle A C B is said to be an angle of forty degrees, and to be measured by A B. But this arc is the same part of the smaller circle that E F is of the greater. The arc A B, therefore, contains the same number of degrees as the arc E F, and either may be taken as the measure of the angle A C B. As the whole circle contains three hundred and sixty degrees, it is evident, that the quarter of a circle, or *quadrant*, contains ninety degrees, and that the semicircle A B D G contains one hundred and eighty degrees.

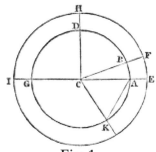

Fig. 1.

The *complement* of an arc, or angle, is what it wants of ninety degrees. Thus, since A D is an arc of ninety degrees, B D is the complement of A B, and A B is the complement of B D. If A B denotes a certain number of degrees of latitude, B D will be the complement of the latitude, or the colatitude, as it is commonly written.

The *supplement* of an arc, or angle, is what it wants of one hundred and eighty degrees. Thus, B A is the supplement of G D B, and G D B is the supplement of B A. If B A were twenty degrees of longitude, G D B, its supplement, would be one hundred and sixty degrees. An angle is said to be *subtended* by the side which is opposite to it. Thus, in the triangle A C K, the angle at C is subtended by the side A K, the angle at A by C K, and the angle at K by C A. In like manner, a side is said to be subtended by an angle, as A K by the angle at C.

Let us now proceed with the doctrine of the sphere.

A section of a sphere, by a plane cutting it in any manner, is a circle. *Great circles* are those which pass through the centre of the sphere, and divide it into two equal hemispheres. *Small circles* are such as do not pass through the centre, but divide the sphere into two unequal parts. The *axis* of a circle is a straight line passing through its centre at right angles to its plane. The *pole* of a great circle is the point on the sphere where its axis cuts through the sphere. Every great circle has two poles, each of which is every where ninety degrees from the great circle. All great circles of the sphere cut each other in two points diametrically opposite, and consequently their points of section are one hundred and eighty degrees apart. A great circle, which passes through the pole of another great circle, cuts the latter at right angles. The great circle which passes through the pole of another great circle, and is at right angles to it, is called a *secondary* to that circle. The angle made by two great circles on the surface of the sphere is measured by an arc of another great circle, of which the angular point is the pole, being the arc of that great circle intercepted between those two circles.

In order to fix the position of any place, either on the surface of the earth or in the heavens, both the earth and the heavens are conceived to be divided into separate portions, by circles, which are imagined to cut through them, in various ways. The earth thus intersected is called the *terrestrial,* and the heavens the *celestial,* sphere. We must bear in mind, that these circles have no existence in Nature, but are mere landmarks, artificially contrived for convenience of reference. On account of the immense distances of the heavenly bodies, they appear to us, wherever we are placed, to be fixed in the same concave surface, or celestial vault. The great circles of the globe, extended every way to meet the concave sphere of the heavens, become circles of the celestial sphere.

The *horizon* is the great circle which divides the earth into upper and lower hemispheres, and separates the visible heavens from the invisible. This is the *rational* horizon. The *sensible* horizon is a circle touching the earth at the place of the spectator, and is bounded by the line in which the earth and skies seem to meet. The sensible horizon is parallel to the rational, but is distant from it by the semidiameter of the earth, or nearly four thousand miles. Still, so vast is the distance of the starry sphere, that both these planes appear to cut the sphere in the same line; so that we see the same hemisphere of stars that we should see, if the upper half of the earth were removed, and we stood on the rational horizon.

The poles of the horizon are the zenith and nadir. The *zenith* is the point directly over our heads; and the *nadir,* that directly under our feet. The plumb-line (such as is formed by suspending a bullet by a string) is in the axis of the horizon, and consequently directed towards its poles. Every place on the surface of the earth has its own horizon; and the traveller has a new horizon at every step, always extending ninety degrees from him, in all directions.

Vertical circles are those which pass through the poles of the horizon, (the zenith and nadir,) perpendicular to it.

The *meridian* is that vertical circle which passes through the north and south points.

The *prime vertical* is that vertical circle which passes through the east and west points.

The *altitude* of a body is its elevation above the horizon, measured on a vertical circle.

The *azimuth* of a body is its distance, measured on the horizon, from the meridian to a vertical circle passing through that body.

The *amplitude* of a body is its distance, on the horizon, from the prime vertical to a vertical circle passing through the body.

Azimuth is reckoned ninety degrees from either the north or south point; and amplitude ninety degrees from either the east or west point. Azimuth and amplitude are mutually complements of each other, for one makes up what the other wants of ninety degrees. When a point is *on* the horizon, it is only necessary to count the number of degrees of the horizon between that point and the meridian, in order to find its azimuth; but if the point is *above* the horizon, then its azimuth is estimated by passing a vertical circle through it, and reckoning the azimuth from the point where this circle cuts the horizon.

The *zenith distance* of a body is measured on a vertical circle passing through that body. It is the complement of the altitude.

The *axis of the earth* is the diameter on which the earth is conceived to turn in its diurnal revolution. The same line, continued until it meets the starry concave, constitutes the *axis of the celestial sphere.*

The *poles of the earth* are the extremities of the earth's axis: the *poles of the heavens*, the extremities of the celestial axis.

The *equator* is a great circle cutting the axis of the earth at right angles. Hence, the axis of the earth is the axis of the equator, and its poles are the poles of the equator. The intersection of the plane of the equator with the surface of the earth constitutes the *terrestrial*, and its intersection with the concave sphere of the heavens, the *celestial*, equator. The latter, by way of distinction, is sometimes denominated the *equinoctial*.

The secondaries to the equator,—that is, the great circles passing through the poles of the equator,—are called *meridians*, because that secondary which passes through the zenith of any place is the meridian of that place, and is at right angles both to the equator and the horizon, passing, as it does, through the poles of both. These secondaries are also called *hour circles* because the arcs of the equator intercepted between them are used as measures of time.

The *latitude* of a place on the earth is its distance from the equator north or south. The *polar distance*, or angular distance from the nearest pole, is the complement of the latitude.

The *longitude* of a place is its distance from some standard meridian, either east or west, measured on the equator. The meridian, usually taken as the standard, is that of the Observatory of Greenwich, in London. If a place is directly *on* the equator, we have only to inquire, how many degrees of the equator there are between that place and the point where the meridian of Greenwich cuts the equator. If the place is north or south of the equator, then its longitude is the arc of the equator intercepted between the meridian which passes through the place and the meridian of Greenwich.

The *ecliptic* is a great circle, in which the earth performs its annual revolutions around the sun. It passes through the centre of the earth and the centre of the sun. It is found, by observation, that the earth does not lie with its axis at right angles to the plane of the ecliptic, so as to make the equator coincide with it, but that it is turned about twenty-three and a half degrees out of a perpendicular direction, making an angle with the plane itself of sixty-six and a half degrees. The equator, therefore, must be turned the same distance out of a coincidence with the ecliptic, the two circles making an angle with each other of twenty-three and a half degrees. It is particularly important that we should form correct ideas of the ecliptic, and of its relations to the equator, since to these two circles a great number of astronomical measurements and phenomena are referred.

The *equinoctial points*, or *equinoxes*, are the intersections of the ecliptic and equator. The time when the sun crosses the equator, in going northward, is called the *vernal*, and in returning southward, the *autumnal*, equinox. The vernal equinox occurs about the twenty-first of March, and the autumnal, about the twenty-second of September.

The *solstitial points* are the two points of the ecliptic most distant from the equator. The times when the sun comes to them are called *solstices*. The Summer solstice occurs about the twenty-second of June, and the Winter solstice about the twenty-second of December. The ecliptic is divided into twelve equal parts, of thirty degrees each, called *signs*, which, beginning at the vernal equinox, succeed each other, in the following order:

1. Aries, ♈	7. Libra, ♎
2. Taurus, ♉	8. Scorpio, ♏
3. Gemini, ♊	9. Sagittarius, ♐
4. Cancer, ♋	10. Capricornus, ♑
5. Leo, ♌	11. Aquarius, ♒
6. Virgo, ♍	12. Pisces. ♓

The mode of reckoning on the ecliptic is by signs, degrees, minutes, and seconds. The sign is denoted either by its name or its number. Thus, one hundred degrees may be expressed either as the tenth degree of Cancer, or as 3s 10°. It will be found an advantage to repeat the signs in their proper order, until they are well fixed in the memory, and to be able to recognise each sign by its appropriate character.

Of the various meridians, two are distinguished by the name of *colures*. The *equinoctial colure* is the meridian which passes through the equinoctial points. From this meridian, right ascension and celestial longitude are reckoned, as longitude on the earth is reckoned from the meridian of Greenwich. The *solstitial colure* is the meridian which passes through the solstitial points.

The position of a celestial body is referred to the equator by its right ascension and declination. *Right ascension* is the angular distance from the vernal equinox measured on the equator. If a star is situated *on* the equator, then its right ascension is the number of degrees of the equator between the star and the vernal equinox. But if the star is north or south of the equator, then its right ascension is the number of degrees of the equator, intercepted between the vernal equinox and that secondary to the equator which passes through the star. *Declination* is the distance of a body from the equator measured on a secondary to the latter. Therefore, right ascension and declination correspond to terrestrial longitude and latitude,—right ascension being reckoned from the equinoctial colure, in the same manner as longitude is reckoned from the meridian of Greenwich. On the other hand, celestial longitude and latitude are referred, not to the equator, but to the ecliptic. *Celestial longitude* is the distance of a body from the vernal equinox measured on the ecliptic. *Celestial latitude* is the distance from the ecliptic measured on a secondary to the latter. Or, more briefly, longitude is distance *on* the ecliptic: latitude, distance *from* the ecliptic. The *north polar distance* of a star is the complement of its declination.

Parallels of latitude are small circles parallel to the equator. They constantly diminish in size, as we go from the equator to the pole. The *tropics* are the parallels of latitude which pass through the solstices. The northern tropic is called the tropic of Cancer; the southern, the tropic of Capricorn. The *polar circles* are the parallels of latitude that pass through the poles of the ecliptic, at the distance of twenty-three and a half degrees from the poles of the earth.

The *elevation of the pole* of the heavens above the horizon of any place is always equal to the latitude of the place. Thus, in forty degrees of north latitude we see the north star forty degrees above the northern horizon; whereas, if we should travel southward, its elevation would grow less and less, until we reached the equator, where it would appear *in* the horizon. Or, if we should travel northwards, the north star would rise continually higher and higher, until, if we could reach the pole of the earth, that star would appear directly over head. The *elevation of the equator* above the horizon of any place is equal to the complement of the latitude. Thus, at the latitude of forty degrees north, the equator is elevated fifty degrees above the southern horizon.

The earth is divided into five zones. That portion of the earth which lies between the tropics is called the *torrid zone*; that between the tropics and the polar circles, the *temperate zones*; and that between the polar circles and the poles, the *frigid zones*.

The *zodiac* is the part of the celestial sphere which lies about eight degrees on each side of the ecliptic. This portion of the heavens is thus marked off by itself, because all the planets move within it.

After endeavoring to form, from the definitions, as clear an idea as we can of the various circles of the sphere, we may next resort to an artificial globe, and see how they are severally represented there. I do not advise to *begin* learning the definitions from the globe; the mind is more improved, and a power of conceiving clearly how things are in Nature is more effectually acquired, by referring every thing, at first, to the grand sphere of

Nature itself, and afterwards resorting to artificial representations to aid our conceptions. We can get but a very imperfect idea of a man from a profile cut in paper, unless we know the original. If we are acquainted with the individual, the profile will assist us to recall his appearance more distinctly than we can do without it. In like manner, orreries, globes, and other artificial aids, will be found very useful, in assisting us to form distinct conceptions of the relations existing between the different circles of the sphere, and of the arrangements of the heavenly bodies; but, unless we have already acquired some correct ideas of these things, by contemplating them as they are in Nature, artificial globes, and especially orreries, will be apt to mislead us.

I trust you will be able to obtain the use of a globe,[1] to aid you in the study of the foregoing definitions, or doctrine of the sphere; but if not, I would recommend the following easy device. To represent the earth, select a large *apple*, (a melon, when in season, will be found still better.) The eye and the stem of the apple will indicate the position of the two poles of the earth. Applying the thumb and finger of the left hand to the poles, and holding the apple so that the poles may be in a north and south line, turn this globe from west to east, and its motion will correspond to the diurnal movement of the earth. Pass a wire or a knitting needle through the poles, and it will represent the *axis* of the sphere. A circle cut around the apple, half way between the poles, will be the *equator*; and several other circles cut between the equator and the poles, parallel to the equator, will represent *parallels of latitude*; of which, two, drawn twenty-three and a half degrees from the equator, will be the *tropics*, and two others, at the same distance from the poles, will be the *polar circles*. A great circle cut through the poles, in a north and south direction, will form the *meridian*, and several other great circles drawn through the poles, and of course perpendicularly to the equator, will be secondaries to the equator, constituting meridians, or *hour circles*. A great circle cut through the centre of the earth, from one tropic to the other, would represent the *plane* of the ecliptic; and consequently a line cut round the apple where such a section meets the surface, will be the terrestrial *ecliptic*. The points where this circle meets the tropics indicate the position of the *solstices*; and its intersection with the equator, that of the *equinoctial points*.

The *horizon* is best represented by a circular piece of pasteboard, cut so as to fit closely to the apple, being movable upon it. When this horizon is passed through the poles, it becomes the horizon of the equator; when it is so placed as to coincide with the earth's equator, it becomes the horizon of the poles; and in every other situation it represents the horizon of a place on the globe ninety degrees every way from it. Suppose we are in latitude forty degrees; then let us place our movable paper parallel to our own horizon, and elevate the pole forty degrees above it, as near as we can judge by the eye. If we cut a circle around the apple, passing through its highest part, and through the east and west points, it will represent the *prime vertical.*

Simple as the foregoing device is, if you will take the trouble to construct one for yourself, it will lead you to more correct views of the doctrine of the sphere, than you would be apt to obtain from the most expensive artificial globes, although there are many other useful purposes which such globes serve, for which the apple would be inadequate. When you have thus made a sphere for yourself, or, with an artificial globe before you, if you have access to one, proceed to point out on it the various arcs of azimuth and altitude, right ascension and declination, terrestrial and celestial latitude and longitude,—these last being referred to the equator on the earth, and to the ecliptic in the heavens.

When the circles of the sphere are well learned, we may advantageously employ projections of them in various illustrations. By the *projection of the sphere* is meant a representation of all its parts on a plane. The plane itself is called the plane of projection. Let us take any circular ring, as a wire bent into a circle, and hold it in different positions before the eye. If we hold it parallel to the face, with the whole breadth opposite to the eye, we see it as an entire circle. If we turn it a little sideways, it appears oval, or as an ellipse; and, as we continue to turn it more and more round, the ellipse grows narrower and narrower, until, when the edge is presented to the eye, we see nothing but a line. Now imagine the ring to be near a perpendicular wall, and the eye to be removed at such a distance from it, as not to distinguish any interval between the ring and the wall; then the several figures under which the ring is seen will appear to be inscribed on the wall, and we shall see the ring as a circle, when perpendicular to a straight line joining the centre of the ring and the eye, or as an ellipse, when oblique to this line, or as a straight line, when its edge is towards us.

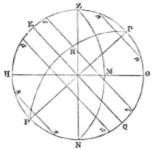

Fig. 2.

It is in this manner that the circles of the sphere are projected, as represented in the following diagram, Fig. 2. Here, various circles are represented as projected on the meridian, which is supposed to be situated directly before the eye, at some distance from it. The horizon H O, being perpendicular to the meridian, is seen edgewise, and consequently is projected into a straight line. The same is the case with the prime vertical Z N, with the equator E Q, and the several small circles parallel to the equator, which represent the two tropics and the two polar circles. In fact, all circles whatsoever, which are perpendicular to the plane of projection, will be represented by straight lines. But every circle which is perpendicular to the horizon, except the prime vertical, being seen obliquely, as Z M N, will be projected into an ellipse, one half only of which is seen,—the other half being on the other side of the plane of projection. In the same manner, P R P, an hour circle, is represented by an ellipse on the plane of projection.

LETTER III.

ASTRONOMICAL INSTRUMENTS.——TELESCOPE.

"Here truths sublime, and sacred science charm,
Creative arts new faculties supply,
Mechanic powers give more than giant's arm,
And piercing optics more than eagle's eye;
Eyes that explore creation's wondrous laws,
And teach us to adore the great Designing Cause."—*Beattie.*

IF, as I trust, you have gained a clear and familiar knowledge of the circles and divisions of the sphere, and of the mode of estimating the position of a heavenly body by its azimuth and altitude, or by its right ascension and declination, or by its longitude and latitude, you will now enter with advantage upon an account of those *instruments*, by means of which our knowledge of astronomy has been greatly promoted and perfected.

The most ancient astronomers employed no instruments of observation, but acquired their knowledge of the heavenly bodies by long-continued and most attentive inspection with the naked eye. Instruments for measuring angles were first used in the Alexandrian school, about three hundred years before the Christian era.

Wherever we are situated on the earth, we appear to be in the centre of a vast sphere, on the concave surface of which all celestial objects are inscribed. If we take any two points on the surface of the sphere, as two stars, for example, and imagine straight lines to be drawn to them from the eye, the angle included between these lines will be measured by the arc of the sky contained between the two points. Thus, if D B H, Fig. 3, page 30, represents the concave surface of the sphere, A, B, two points on it, as two stars, and C A, C B, straight lines drawn from the spectator to those points, then the angular distance between them is measured by the arc A B, or the angle A C B. But this angle may be measured on a much smaller circle, having the same centre, as G F K, since the arc E F will have the same number of degrees as the arc A B. The simplest mode of taking an angle between two stars is by means of an arm opening at a joint like the blade of a penknife, the end of the arm moving like C E upon the graduated circle K F G. In fact, an instrument constructed on this principle, resembling a carpenter's rule with a folding joint, with a semicircle attached, constituted the first rude apparatus for measuring the angular distance between two points on the celestial sphere. Thus the sun's elevation above the horizon might be ascertained, by placing one arm of the rule on a level with the horizon, and bringing the edge of the other into a line with the sun's centre.

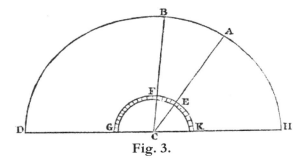

Fig. 3.

The common surveyor's compass affords a simple example of angular measurement. Here, the needle lies in a north and south line, while the circular rim of the compass, when the instrument is level, corresponds to the horizon. Hence the compass shows the azimuth of an object, or how many degrees it lies east or west of the meridian.

It is obvious, that the larger the graduated circle is, the more minutely its limb may be divided. If the circle is one foot in diameter, each degree will occupy one tenth of an inch. If the circle is twenty feet in diameter, a degree will occupy the space of two inches, and could be easily divided into minutes, since each minute would cover a space one thirtieth of an inch. Refined astronomical circles are now divided with very great skill and accuracy, the spaces between the divisions being, when read off, magnified by a microscope; but in former times, astronomers had no mode of measuring small angles but by employing very large circles. But the telescope and microscope enable us at present to measure celestial arcs much more accurately than was done by the older astronomers. In the best instruments, the measurements extend to a single second of space, or one thirty-six hundredth part of a degree,—a space, on a circle twelve feet in diameter, no greater than one fifty-seven hundredth part of an inch. To divide, or *graduate*, astronomical instruments, to such a degree of nicety, requires the highest efforts of mechanical skill. Indeed, the whole art of instrument-making is regarded as the most difficult and refined of all the mechanical arts; and a few eminent artists, who have produced instruments of peculiar power and accuracy, take rank with astronomers of the highest celebrity.

I will endeavor to make you acquainted with several of the principal instruments employed in astronomical observations, but especially with the telescope, which is the most important and interesting of them all. I think I shall consult your wishes, as well as your improvement, by giving you a clear insight into the principles of this prince of instruments, and by reciting a few particulars, at least, respecting its invention and subsequent history.

The *Telescope*, as its name implies, is an instrument employed for viewing distant objects.[2] It aids the eye in two ways; first, by enlarging the visual angle under which objects are seen, and, secondly, by collecting and conveying to the eye a much larger amount of the light that emanates from the object, than would enter the naked pupil. A complete knowledge of the telescope cannot be acquired, without an acquaintance with the science of optics; but one unacquainted with that science may obtain some idea of the leading principles of this noble instrument. Its main principle is as follows: *By means of the telescope, we first form an image of a distant object,—as the moon, for example,—and then magnify that image by a microscope.*

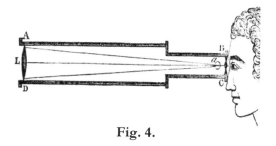

Fig. 4.

Let us first see how the image is formed. This may be done either by a convex lens, or by a concave mirror. A convex lens is a flat piece of glass, having its two faces convex, or spherical, as is seen in a common sun-glass, or a pair of spectacles. Every one who has seen a sun-glass, knows, that, when held towards the sun, it collects the solar rays into a small bright circle in the focus. This is in fact a small *image* of the sun. In the same manner, the image of any distant object, as a star, may be formed, as is represented in the following diagram. Let A B C D, Fig. 4, represent the tube of the telescope. At the front end, or at the end which is directed towards the object, (which we will suppose to be the moon,) is inserted a convex lens, L, which receives the rays of light from the moon, and collects them into the focus at *a*, forming an image of the moon. This image is viewed by a magnifier attached to the end B C. The lens, L, is called the *object-glass*, and the microscope in B C, the *eyeglass*. We apply a microscope to this image just as we would to any object; and, by greatly enlarging its dimensions, we may render its various parts far more distinct than they would otherwise be; while, at the same time, the lens collects and conveys to the eye a much greater quantity of light than would proceed directly from the body under examination. A very few rays of light only, from a distant object, as a star, can enter the eye directly; but a lens one foot in diameter will collect a beam of light of the same dimensions, and convey it to the eye. By these means, many obscure celestial objects become distinctly visible, which would otherwise be either too minute, or not sufficiently luminous, to be seen by us.

But the image may also be formed by means of a *concave mirror*, which, as well as the concave lens, has the property of collecting the rays of light which proceed from any luminous body, and of forming an image of that body. The image formed by a concave mirror is magnified by a microscope, in the same manner as when formed by the concave lens. When the lens is used to form an image, the instrument is called a *refracting telescope*; when a concave mirror is used, it is called a *reflecting telescope*.

The office of the object-glass is simply *to collect* the light, and to form an *image* of the object, but not to magnify it: the magnifying power is wholly in the eyeglass. Hence the principle of the telescope is as follows: *By means of the object-glass, (in the refracting telescope,) or by the concave mirror, (in the reflecting telescope,) we form an image of the object, and magnify that image by a microscope.*

The invention of this noble instrument is generally ascribed to the great philosopher of Florence, Galileo. He had heard that a spectacle maker of Holland had accidentally hit upon a discovery, by which distant objects might be brought apparently nearer; and, without further information, he pursued the inquiry, in order to ascertain what forms and combinations of glasses would produce such a result. By a very philosophical process of reasoning, he was led to the discovery of that peculiar form of the telescope which bears his name.

Although the telescopes made by Galileo were no larger than a common spyglass of the kind now used on board of ships, yet, as they gave new views of the heavenly bodies, revealing the mountains and valleys of the moon, the satellites of Jupiter, and multitudes of stars which are invisible to the naked eye, it was regarded with infinite delight and astonishment.

Reflecting telescopes were first constructed by Sir Isaac Newton, although the use of a concave reflector, instead of an object-glass, to form the image, had been previously suggested by Gregory, an eminent Scotch astronomer. The first telescope made by Newton was only six inches long. Its reflector, too, was only a little more than an inch. Notwithstanding its small dimensions, it performed so well, as to encourage further efforts; and this illustrious philosopher afterwards constructed much larger instruments, one of which, made with his own hands, was presented to the Royal Society of London, and is now carefully preserved in their library.

Newton was induced to undertake the construction of reflecting telescopes, from the belief that refracting telescopes were necessarily limited to a very small size, with only moderate illuminating powers, whereas the dimensions and powers of the former admitted of being indefinitely increased. Considerable *magnifying* powers might, indeed, be obtained from refractors, by making them very long; but the *brightness* with which telescopic objects are seen, depends greatly on the dimensions of the beam of light which is collected by the object-glass, or by the mirror, and conveyed to the eye; and therefore, small object-glasses cannot have a very high illuminating power. Now, the experiments of Newton on colors led him to believe, that it would be impossible to employ large lenses in the construction of telescopes, since such glasses would give to the images, they formed, the colors of the rainbow. But later opticians have found means of correcting these imperfections, so that we are now able to use object-glasses a foot or more in diameter, which give very clear and bright images. Such instruments are called *achromatic* telescopes,—a name implying the absence of prismatic or rainbow colors in the image. It is, however, far more difficult to construct large achromatic than large reflecting telescopes. Very large pieces of glass can seldom be found, that are sufficiently pure for the purpose; since every inequality in the glass, such as waves, tears, threads, and the like, spoils it for optical purposes, as it distorts the light, and produces nothing but confused images.

The achromatic telescope (that is, the refracting telescope, having such an object-glass as to give a colorless image) was invented by Dollond, a distinguished English artist, about the year 1757. He had in his possession a quantity of glass of a remarkably fine quality, which enabled him to carry his invention at once to a high degree of perfection. It has ever since been, with the manufacturers of telescopes, a matter of the greatest difficulty to find pieces of glass, of a suitable quality for object-glasses, more than two or three inches in diameter. Hence, large achromatic telescopes are very expensive, being valued in proportion to the *cubes* of their diameters; that is, if a telescope whose aperture (as the breadth of the object-glass is technically called) is two inches, cost one hundred dollars, one whose aperture is eight inches would cost six thousand four hundred dollars.

Since it is so much easier to make large reflecting than large refracting telescopes, you may ask, why the latter are ever attempted, and why reflectors are not exclusively employed? I answer, that the achromatic telescope, when large and well constructed, is a more perfect and more durable instrument than the reflecting telescope. Much more of the light that falls on the mirror is absorbed than is lost in passing through the object-glass of a refractor; and hence the larger achromatic telescopes afford a stronger light than the reflecting, unless the latter are made of an enormous and unwieldy size. Moreover, the mirror is very liable to tarnish, and will never retain its full lustre for many years together; and it is no easy matter to restore the lustre, when once impaired.

In my next Letter, I will give you an account of some of the most celebrated telescopes that have ever been constructed, and point out the method of using this excellent instrument, so as to obtain with it the finest views of the heavenly bodies.

LETTER IV

TELESCOPE CONTINUED.

——"the broad circumference
 Hung on his shoulders like the moon, whose orb
Through *optic glass* the Tuscan artist views
vening, from the top of Fesolé
Or in Valdarno, to descry new lands,
Rivers or mountains, in her spotted globe."—*Milton.*

THE two most celebrated telescopes, hitherto made, are Herschel's *forty-feet reflector*, and the *great Dorpat refractor.* Herschel was a Hanoverian by birth, but settled in England in the younger part of his life. As early as 1774, he began to make telescopes for his own use; and, during his life, he made more than four hundred, of various sizes and powers. Under the patronage of George the Third, he completed, in 1789, his great telescope, having a tube of iron, forty feet long, and a speculum, forty-nine and a half inches or more than four feet in diameter. Let us endeavor to form a just conception of this gigantic instrument, which we can do only by dwelling on its dimensions, and comparing them with those of other objects with which we are familiar, as the length or height of a house, and the breadth of a hogshead or cistern, of known dimensions. The reflector alone weighed nearly a ton. So large and ponderous an instrument must require a vast deal of machinery to work it, and to keep it steady; and, accordingly, the framework surrounding it was formed of heavy timbers, and resembled the frame of a large building. When one of the largest of the fixed stars, as Sirius, is entering the field of this telescope, its approach is announced by a bright dawn, like that which precedes the rising sun; and when the star itself enters the field, the light is insupportable to the naked eye. The planets are expanded into brilliant luminaries, like the moon; and innumerable multitudes of stars are scattered like glittering dust over the celestial vault.

The great Dorpat telescope is of more recent construction. It was made by Fraunhofer, a German optician of the greatest eminence, at Munich, in Bavaria, and takes its name from its being attached to the observatory at Dorpat, in Russia. It is of much smaller dimensions than the great telescope of Herschel. Its object-glass is nine and a half inches in diameter, and its length, fourteen feet. Although the price of this instrument was nearly five thousand dollars, yet it is said that this sum barely covered the actual expenses. It weighs five thousand pounds, and yet is turned with the finger. In facility of management, it has greatly the advantage of Herschel's telescope. Moreover, the sky of England is so much of the time unfavorable for astronomical observation, that *one hundred* good hours (or those in which the higher powers can be used) are all that can be obtained in a whole year. On this account, and on account of the difficulty of shifting the position of the instrument, Herschel estimated that it would take about six hundred years to obtain with it even a momentary glimpse of every part of the heavens. This remark shows that such great telescopes are unsuited to the common purposes of astronomical observation. Indeed, most of Herschel's discoveries were made with his small telescopes; and although, for certain rare purposes, powers were applied which magnified seven thousand times, yet, in most of his observations, powers magnifying only two or three hundred times were employed. The highest power of the Dorpat telescope is only seven hundred, and yet the director of this instrument, Professor Struve, is of the opinion, that it is nearly or quite equal in quality, all things considered, to Herschel's forty-feet reflector.

It is not generally understood in what way greatness of size in a telescope increases its powers; and it conveys but an imperfect idea of the excellence of a telescope, to tell how much it magnifies. In the same instrument, an increase of magnifying power is always attended with a diminution of the light and of the field of view. Hence, the lower powers generally afford the most agreeable views, because they give the clearest light, and take in the largest space. The several circumstances which influence the qualities of a telescope are, illuminating power, distinctness, field of view, and magnifying power. Large mirrors and large object-glasses are superior to smaller ones, because they collect a larger beam of light, and transmit it to the eye. Stars which are invisible to the naked eye are rendered visible by the telescope, because this instrument collects and conveys to the eye a large beam of the few rays which emanate from the stars; whereas a beam of these rays of only the diameter of the pupil of the eye, would afford too little light for distinct vision. In this particular, large telescopes have great advantages over small ones. The great mirror of Herschel's forty-feet reflector collects and conveys to the eye a beam more than four feet in diameter. The Dorpat telescope also transmits to the eye a beam nine and one half inches in diameter. This seems small, in comparison with the reflector; but much less of the light is lost on passing through the glass than is absorbed by the mirror, and the mirror is very liable to be clouded or tarnished; so that there is not so great a difference in the two instruments, in regard to illuminating power, as might be supposed from the difference of size.

Distinctness of view is all-important to the performance of an instrument. The object may be sufficiently bright, yet, if the image is distorted, or ill-defined, the illumination is of little consequence. This property depends mainly on the skill with which all the imperfections of figure and color in the glass or mirror are corrected, and can exist in perfection only when the image is rendered completely achromatic, and when all the rays that proceed from each point in the object are collected into corresponding points of the image, unaccompanied by any other rays. Distinctness is very much affected by the *steadiness* of the instrument. Every one knows how indistinct a page becomes, when a book is passed rapidly backwards and forwards before the eyes, and how difficult it is to read in a carriage in rapid motion on a rough road.

Field of view is another important consideration. The finest instruments exhibit the moon, for example, not only bright and distinct, in all its parts, but they take in the whole disk at once; whereas, the inferior instruments, when the higher powers, especially, are applied, permit us to see only a small part of the moon at once.

I hope, my friend, that, when you have perused these Letters, or rather, while you are perusing them, you will have frequent opportunities of looking through a good telescope. I even anticipate that you will acquire such a taste for viewing the heavenly bodies with the aid of a good glass, that you will deem a telescope a most suitable appendage to your library, and as certainly not less an ornament to it than the more expensive statues with which some people of fortune adorn theirs. I will therefore, before concluding this letter, offer you a few *directions for using the telescope.*

Some states of weather, even when the sky is clear, are far more favorable for astronomical observation than others. After sudden changes of temperature in the atmosphere, the medium is usually very unsteady. If the sun shines out warm after a cloudy season, the ground first becomes heated, and the air that is nearest to it is expanded, and rises, while the colder air descends, and thus ascending and descending currents of air, mingling together, create a confused and wavy medium. The same cause operates when a current of hot air rises from a chimney; and hence the state of the atmosphere in cities and large towns is very unfavorable to the astronomer, on this account, as well as on account of the smoky condition in which it is usually found. After a long season of dry weather, also, the air becomes smoky, and unfit for observation. Indeed, foggy, misty, or smoky, air is so prevalent in some countries, that only a very few times in the whole year can be found, which are entirely suited to observation, especially with the higher powers; for we must recollect, that these inequalities and imperfections are magnified by telescopes, as well as the objects themselves. Thus, as I have already mentioned, not more than one hundred good hours in a year could be obtained for observation with Herschel's great telescope. By *good* hours, Herschel means that the sky must be very clear, the moon absent, no twilight, no haziness, no violent wind, and no sudden change of temperature. As a general fact, the warmer climates enjoy a much finer sky for the astronomer than the colder, having many more clear evenings, a short twilight, and less change of temperature. The watery vapor of the atmosphere, also, is more perfectly dissolved in hot than in cold air, and the more water air contains, provided it is in a state of perfect solution, the clearer it is.

A *certain preparation of the observer himself* is also requisite for the nicest observations with the telescope. He must be free from all agitation, and the eye must not recently have been exposed to a strong light, which contracts the pupil of the eye. Indeed, for delicate observations, the observer should remain for some time beforehand in a dark room, to let the pupil of the eye dilate. By this means, it will be enabled to admit a larger number of the rays of light. In ascending the stairs of an observatory, visitors frequently get out of breath, and having perhaps recently emerged from a strongly-lighted apartment, the eye is not in a favorable state for observation. Under these disadvantages, they take a hasty look into the telescope, and it is no wonder that disappointment usually follows.

Want of steadiness is a great difficulty attending the use of the highest magnifiers; for the motions of the instrument are magnified as well as the object. Hence, in the structure of observatories, the greatest pains is requisite, to avoid all tremor, and to give to the instruments all possible steadiness; and the same care is to be exercised by observers. In the more refined observations, only one or two persons ought to be near the instrument.

In general, *low powers* afford better views of the heavenly bodies than very high magnifiers. It may be thought absurd, to recommend the use of low powers, in respect to large instruments especially, since it is commonly supposed that the advantage of large instruments is, that they will bear high magnifying powers. But this is not their only, nor even their principal, advantage. A good light and large field are qualities, for most purposes, more important than great magnifying power; and it must be borne in mind, that, as we increase the magnifying power in a given instrument, we diminish both the illumination and the field of view. Still, different objects require different magnifying powers; and a telescope is usually furnished with several varieties of powers, one of which is best fitted for viewing the moon, another for Jupiter, and a still higher power for Saturn. Comets require only the lowest magnifiers; for here, our object is to command as much light, and as large a field, as possible, while it avails little to increase the dimensions of the object. On the other hand, for certain double stars, (stars which appear single to the naked eye, but double to the telescope,) we require very high magnifiers, in order to separate these minute objects so far from each other, that the interval can be distinctly seen. Whenever we exhibit celestial objects to inexperienced observers, it is useful to precede the view with good *drawings* of the objects, accompanied by an explanation of what each appearance, exhibited in the telescope, indicates. The novice is told, that mountains and valleys can be seen in the moon by the aid of the telescope; but, on looking, he sees a confused mass of light and shade, and nothing which looks to him like either mountains or valleys. Had his attention been previously directed to a plain drawing of the moon, and each particular appearance interpreted to him, he would then have looked through the telescope with intelligence and satisfaction.

LETTER V.

OBSERVATORIES.

"We, though from heaven remote, to heaven will move,
With strength of mind, and tread the abyss above;
And penetrate, with an interior light,
Those upper depths which Nature hid from sight.
Pleased we will be, to walk along the sphere
Of shining stars, and travel with the year."—*Ovid.*

AN observatory is a structure fitted up expressly for astronomical observations, and furnished with suitable instruments for that purpose.

The two most celebrated observatories, hitherto built, are that of Tycho Brahe, and that of Greenwich, near London. The observatory of Tycho Brahe, Fig. 5, was constructed at the expense of the King of Denmark, in a style of royal magnificence, and cost no less than two hundred thousand crowns. It was situated on the island of Huenna, at the entrance of the Baltic, and was called Uraniburg, or the palace of the skies.

Before I give you an account of Tycho's observatory, I will recite a few particulars respecting this great astronomer himself.

Fig. 5.

Tycho Brahe was of Swedish descent, and of noble family; but having received his education at the University of Copenhagen, and spent a large part of his life in Denmark, he is usually considered as a Dane, and quoted as a Danish astronomer. He was born in the year 1546. When he was about fourteen years old, there happened a great eclipse of the sun, which awakened in him a high interest, especially when he saw how accurately all the circumstances of it answered to the prediction with which he had been before made acquainted. He was immediately seized with an irresistible passion to acquire a knowledge of the science which could so successfully lift the veil of futurity. His friends had destined him for the profession of law, and, from the superior talents of which he gave early promise, and with the advantage of powerful family connexions, they had marked out for him a distinguished career in public life. They therefore endeavored to discourage him from pursuing a path which they deemed so much less glorious than that, and vainly sought, by various means, to extinguish the zeal for astronomy which was kindled in his youthful bosom. Despising all the attractions of a court, he contracted an alliance with a peasant girl, and, in the peaceful retirement of domestic life, desired no happier lot than to peruse the grand volume which the nocturnal heavens displayed to his enthusiastic imagination. He soon established his fame as one of the greatest astronomers of the age, and monarchs did homage to his genius. The King of Denmark became his munificent patron, and James the First, King of England, when he went to Denmark to complete his marriage with a Danish Princess, passed eight days with Tycho in his observatory, and, at his departure, addressed to the astronomer a Latin ode, accompanied with a magnificent present. He gave him also his royal license to print his works in England, and added to it the following complimentary letter: "Nor am I acquainted with these things on the relation of others, or from a mere perusal of your works, but I have seen them with my own eyes, and heard them with my own ears, in your residence at Uraniburg, during the various learned and agreeable conversations which I there held with you, which even now affect my mind to such a degree, that it is difficult to decide, whether I recollect them with greater pleasure or admiration." Admiring disciples also crowded to this sanctuary of the sciences, to acquire a knowledge of the heavens.

The observatory consisted of a main building, which was square, each side being sixty feet, and of large wings in the form of round towers. The whole was executed in a style of great magnificence, and Tycho, who was a nobleman by descent, gratified his taste for splendor and ornament, by giving to every part of the structure an air of the most finished elegance. Nor were the instruments with which it was furnished less magnificent than the buildings. They were vastly larger than had before been employed in the survey of the heavens, and many of them were adorned with costly ornaments. The cut on page 46, Fig. 6, represents one of Tycho's large and splendid instruments, (an astronomical quadrant,) on one side of which was figured a representation of the astronomer and his assistants, in the midst of their instruments, and intently engaged in making and recording observations. It conveys to us a striking idea of the magnificence of his arrangements, and of the extent of his operations.

Here Tycho sat in state, clad in the robes of nobility, and supported throughout his establishment the etiquette due to his rank. His observations were more numerous than all that had ever been made before, and they were carried to a degree of accuracy that is astonishing, when we consider that they were made without the use of the telescope, which was not yet invented.

Tycho carried on his observations at Uraniburg for about twenty years, during which time he accumulated an immense store of accurate and valuable *facts*, which afforded the groundwork of the discovery of the great laws of the solar system established by Kepler, of whom I shall tell you more hereafter.

Fig. 6.

But the high marks of distinction which Tycho enjoyed, not only from his own Sovereign, but also from foreign potentates, provoked the envy of the courtiers of his royal patron. They did not indeed venture to make their attacks upon him while his generous patron was living; but the King was no sooner dead, and succeeded by a young monarch, who did not feel the same interest in protecting and encouraging this great ornament of the kingdom, than his envious foes carried into execution their long-meditated plot for his ruin. They represented to the young King, that the treasury was exhausted, and that it was necessary to retrench a number of pensions, which had been granted for useless purposes, and in particular that of Tycho, which, they maintained, ought to be conferred upon some person capable of rendering greater services to the state. By these means, they succeeded in depriving him of his support, and he was compelled to retreat under the hospitable mansion of a friend in Germany. Here he became known to the Emperor, who invited him to Prague, where, with an ample stipend, he resumed his labors. But, though surrounded with affectionate friends and admiring disciples, he was still an exile in a foreign land. Although his country had been base in its ingratitude, it was yet the land which he loved; the scene of his earliest affection; the theatre of his scientific glory. These feelings continually preyed upon his mind, and his unsettled spirit was ever hovering among his native mountains. In this condition he was attacked by a disease of the most painful kind, and, though its agonizing paroxysms had lengthened intermissions, yet he saw that death was approaching. He implored his pupils to persevere in their scientific labors; he conversed with Kepler on some of the profoundest points of astronomy; and with these secular occupations he mingled frequent acts of piety and devotion. In this happy condition he expired, without pain, at the age of fifty-five.[3]

The observatory at Greenwich was not built until a hundred years after that of Tycho Brahe, namely, in 1676. The great interests of the British nation, which are involved in navigation, constituted the ruling motive with the government to lend their aid in erecting and maintaining this observatory.

The site of the observatory at Greenwich is on a commanding eminence facing the River Thames, five miles east of the central parts of London. Being part of a royal park, the neighboring grounds are in no danger of being occupied by buildings, so as to obstruct the view. It is also in full view of the shipping on the Thames; and, according to a standing regulation of the observatory, at the instant of one o'clock, every day, a huge ball is dropped from over the house, as a signal to the commanders of vessels for regulating their chronometers.

The buildings comprise a series of rooms, of sufficient number and extent to accommodate the different instruments, the inmates of the establishment, and the library; and on the top is a celebrated camera obscura, exhibiting a most distinct and perfect picture of the grand and unrivalled scenery which this eminence commands.

This establishment, by the accuracy and extent of its observations, has contributed more than all other institutions to perfect the science of astronomy.

To preside over and direct this great institution, a man of the highest eminence in the science is appointed by the government, with the title of *Astronomer Royal*. He is paid an ample salary, with the understanding that he is to devote himself exclusively to the business of the observatory. The astronomers royal of the Greenwich observatory, from the time of its first establishment, in 1676, to the present time, have constituted a series of the proudest names of which British science can boast. A more detailed sketch of their interesting history will be given towards the close of these Letters.

Six assistants, besides inferior laborers, are constantly in attendance; and the business of making and recording observations is conducted with the utmost system and order.

The great objects to be attained in the construction of an observatory are, a commanding and unobstructed view of the heavens; freedom from causes that affect the transparency and uniform state of the atmosphere, such as fires, smoke, or marshy grounds; mechanical facilities for the management of instruments, and, especially, every precaution that is necessary to secure perfect steadiness. This last consideration is one of the greatest importance, particularly in the use of very large magnifiers; for we must recollect, that any motion in the instrument is magnified by the full power of the glass, and gives a proportional unsteadiness to the object. A situation is therefore selected as remote as possible from public roads, (for even the passing of carriages would give a tremulous motion to the ground, which would be sensible in large instruments,) and structures of solid masonry are commenced deep enough in the ground to be unaffected by frost, and built up to the height required, without any connexion with the other parts of the building. Many observatories are furnished with a movable dome for a roof, capable of revolving on rollers, so that instruments penetrating through the roof may be easily brought to bear upon any point at or near the zenith.

You will not perhaps desire me to go into a minute description of all the various instruments that are used in a well-constructed observatory. Nor is this necessary, since a very large proportion of all astronomical observations are taken on the meridian, by means of the transit instrument and clock. When a body, in its diurnal revolution, comes to the meridian, it is at its highest point above the horizon, and is then least affected by refraction and parallax. This, then, is the most favorable position for taking observations upon it. Moreover, it is peculiarly easy to take observations on a body when in this situation. Hence the transit instrument and clock are the most important members of an astronomical observatory. You will, therefore, expect me to give you some account of these instruments.

Fig. 7.

The *transit instrument* is a telescope which is fixed permanently in the meridian, and moves only in that plane. The accompanying diagram, Fig. 7, represents a side view of a portable transit instrument, exhibiting the telescope supported on a firm horizontal axis, on which it turns in the plane of the meridian, from the south point of the horizon through the zenith to the north point. It can therefore be so directed as to observe the passage of a star across the meridian at any altitude. The accompanying graduated circle enables the observer to set the instrument at any required altitude, corresponding to the known altitude at which the body to be observed crosses the meridian. Or it may be used to measure the altitude of a body, or its zenith distance, at the time of its meridian passage. Near the circle may be seen a spirit-level, which serves to show when the axis is exactly on a level with the horizon. The framework is made of solid metal, (usually brass,) every thing being arranged with reference to keeping the instrument perfectly steady. It stands on screws, which not only afford a steady support, but are useful for adjusting the instrument to a perfect level. The transit instrument is sometimes fixed immovably to a solid foundation, as a pillar of stone, which is built up from a depth in the ground below the reach of frost. When enclosed in a building, as in an observatory, the stone pillar is carried up separate from the walls and floors of the building, so as to be entirely free from the agitations to which they are liable.

The use of the transit instrument is to show the precise instant when a heavenly body is on the meridian, or to measure the time it occupies in crossing the meridian. The *astronomical clock* is the constant companion of the transit instrument. This clock is so regulated as to keep exact pace with the stars, and of course with the revolution of the earth on its axis; that is, it is regulated to *sidereal* time. It measures the progress of a star, indicating an hour for every fifteen degrees, and twenty-four hours for the whole period of the revolution of the star. Sidereal time commences when the vernal equinox is on the meridian, just as solar time commences when the sun is on the meridian. Hence the hour by the sidereal clock has no correspondence with the hour of the day, but simply indicates how long it is since the equinoctial point crossed the meridian. For example, the clock of an observatory points to three hours and twenty minutes; this may be in the morning, at noon, or any other time of the day,—for it merely shows that it is three hours and twenty minutes since the equinox was on the meridian. Hence, when a star is on the meridian, the clock itself shows its right ascension, which you will recollect is the angular distance measured on the equinoctial, from the point of intersection of the ecliptic and equinoctial, called the vernal equinox, reckoning fifteen degrees for every hour, and a proportional number of degrees and minutes for a less period. I have before remarked, that a very large portion of all astronomical observations are taken when the bodies are on the meridian, by means of the transit instrument and clock.

Having now described these instruments, I will next explain the manner of using them for different observations. Any thing becomes a measure of time, which divides duration equally. The equinoctial, therefore, is peculiarly adapted to this purpose, since, in the daily revolution of the heavens, equal portions of the equinoctial pass under the meridian in equal times. The only difficulty is, to ascertain the amount of these portions for given intervals. Now, the clock shows us exactly this amount; for, when regulated to sidereal time, (as it easily may be,) the hour-hand keeps exact pace with the equator, revolving once on the dial-plate of the clock while the equator turns once by the revolution of the earth. The same is true, also, of all the small circles of diurnal revolution; they all turn exactly at the same rate as the equinoctial, and a star situated any where between the equator and the pole will move in its diurnal circle along with the clock, in the same manner as though it were in the equinoctial. Hence, if we note the interval of time between the passage of any two stars, as shown by the clock, we have a measure of the number of degrees by which they are distant from each other in right ascension. Hence we see how easy it is to take arcs of right ascension: the transit instrument shows us when a body is on the meridian; the clock indicates how long it is since the vernal equinox passed it, which is the right ascension itself; or it tells us the difference of right ascension between any two bodies, simply by indicating the difference in time between their periods of passing the meridian. Again, it is easy to take the *declination* of a body when on the meridian. By declination, you will recollect, is meant the distance of a heavenly body from the equinoctial; the same, indeed, as latitude on the earth. When a star is passing the meridian, if, on the instant of crossing the meridian wire of the telescope, we take its distance from the north pole, (which may readily be done, because the position of the pole is always known, being equal to the latitude of the place,) and subtract this distance from ninety degrees, the remainder will be the distance from the equator, which is the declination. You will ask, why we take this indirect method of finding the declination? Why we do not rather take the distance of the star from the equinoctial, at once? I answer, that it is easy to point an instrument to the north pole, and to ascertain its exact position, and of course to measure any distance from it on the meridian, while, as there is nothing to mark the exact situation of the equinoctial, it is not so easy to take direct measurements from it. When we have thus determined the situation of a heavenly body, with respect to two great circles at right angles with each other, as in the present case, the distance of a body from the equator and from the equinoctial colure, or that meridian which passes though the vernal equinox, we know its relative position in the heavens; and when we have thus determined the relative positions of all the stars, we may lay them down on a map or a globe, exactly as we do places on the earth, by means of their latitude and longitude.

The foregoing is only a *specimen* of the various uses of the transit instrument, in finding the relative places of the heavenly bodies. Another use of this excellent instrument is, to regulate our clocks and watches. By an observation with the transit instrument, we find when the sun's centre is on the meridian. This is the exact time of *apparent* noon. But watches and clocks usually keep *mean* time, and therefore, in order to set our timepiece by the transit instrument, we must apply to the apparent time of noon the equation of time, as will be explained in my next Letter.

A *noon-mark* may easily be made by the aid of the transit instrument. A window sill is frequently selected as a suitable place for the mark, advantage being taken of the shadow projected upon it by the perpendicular casing of the window. Let an assistant stand, with a rule laid on the line of shadow, and with a knife ready to make the mark, the instant when the observer at the transit instrument announces that the centre of the sun is on the meridian. By a concerted signal, as the stroke of a bell, the inhabitants of a town may all fix a noon-mark from the same observation. If the signal be given on one of the days when apparent time and mean time become equal to each other, as on the twenty-fourth of December, no equation of time is required.

As a noon-mark is convenient for regulating timepieces, I will point out a method of making one, which may be practised without the aid of the telescope. Upon a smooth, level plane, freely exposed to the sun, with a pair of compasses describe a circle. In the centre, where the leg of the compasses stood, erect a perpendicular wire of such a length, that the termination of its shadow shall fall upon the circumference of the circle at some hour before noon, as about ten o'clock. Make a small dot at the point where the end of the shadow falls upon the circle, and do the same where it falls upon it again in the afternoon. Take a point half-way between these two points, and from it draw a line to the centre, and it will be a true meridian line. The direction of this line would be the same, whether it were made in the Summer or in the Winter; but it is expedient to draw it about the fifteenth of June, for then the shadow alters its length most rapidly, and the moment of its crossing the wire will be more definite, than in the Winter. At this time of year, also, the sun and clock agree, or are together, as will be more fully explained in my next Letter; whereas, at other times of the year, the time of noon, as indicated by a common clock, would not agree with that indicated by the sun. If the upper end of the wire is flattened, and a small hole is made in it, through which the sun may shine, the instant when this bright spot falls upon the circle will be better defined than the termination of the shadow.

Another important instrument of the observatory is the *mural circle*. It is a graduated circle, usually of very large size, fixed permanently in the plane of the meridian, and attached firmly to a perpendicular wall; and on its centre is a telescope, which revolves along with it, and is easily brought to bear on any object in any point in the meridian. It is made of large size, sometimes twenty feet in diameter, in order that very small angles may be measured on its limb; for it is obvious that a small angle, as one second, will be a larger space on the limb of an instrument, in proportion as the instrument itself is larger. The vertical circle usually connected with the transit instrument, as in Fig. 7, may indeed be employed for the same purposes as the mural circle, namely, to measure arcs of the meridian, as meridian altitudes, zenith distances, north polar distances, and declinations; but as that circle must necessarily be small, and therefore incapable of measuring very minute angles, the mural circle is particularly useful in measuring these important arcs. It is very difficult to keep so large an instrument perfectly steady; and therefore it is attached to a massive wall of solid masonry, and is hence called a *mural* circle, from a Latin word, (*murus*,) which signifies a wall.

The diagram, Fig. 8, page 56, represents a mural circle fixed to its wall, and ready for observations. It will be seen, that every expedient is employed to give the instrument firmness of parts and steadiness of position. The circle is of solid metal, usually of brass, and it is strengthened by numerous radii, which keep it from warping or bending; and these are made in the form of hollow cones, because that is the figure which unites in the highest degree lightness and strength. On the rim of the instrument, at A, you may observe a microscope. This is attached to a micrometer,—a delicate piece of apparatus, used for reading the minute subdivisions of angles; for, after dividing the limb of the instrument as minutely as possible, it will then be necessary to magnify those divisions with the microscope, and subdivide each of these parts with the micrometer. Thus, if we have a mural circle twenty feet in diameter, and of course nearly sixty-three feet in circumference, since there are twenty-one thousand and six hundred minutes in the whole circle, we shall find, by calculation, that one minute would occupy, on the limb of such an instrument, only about one thirtieth of an inch, and a second, only one eighteen hundredth of an inch. We could not, therefore, hope to carry the actual divisions to a greater degree of minuteness than minutes; but each of these spaces may again be subdivided into seconds by the micrometer.

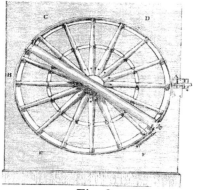

Fig. 8.

From these statements, you will acquire some faint idea of the extreme difficulty of making perfect astronomical instruments, especially where they are intended to measure such minute angles as one second. Indeed, the art of constructing astronomical instruments is one which requires such refined mechanical genius,—so superior a mind to devise, and so delicate a hand to execute,—that the most celebrated instrument-makers take rank with the most distinguished astronomers; supplying, as they do, the means by which only the latter are enabled to make these great discoveries. Astronomers have sometimes made their own telescopes; but they have seldom, if ever, possessed the refined manual skill which is requisite for graduating delicate instruments.

The *sextant* is also one of the most valuable instruments for taking celestial arcs, or the distance between any two points on the celestial sphere, being applicable to a much greater number of purposes than the instruments already described. It is particularly valuable for measuring celestial arcs at sea, because it is not, like most astronomical instruments, affected by the motion of the ship. The principle of the sextant may be briefly described, as follows: it gives the angular distance between any two bodies on the celestial sphere, by reflecting the image of one of the bodies so as to coincide with the other body, as seen directly. The arc through which the reflector is turned, to bring the reflected body to coincide with the other body, becomes a measure of the angular distance between them. By keeping this principle in view, you will be able to understand the use of the several parts of the instrument, as they are exhibited in the diagram, Fig. 9, page 58.

It is, you observe, of a triangular shape, and it is made strong and firm by metallic cross-bars. It has two reflectors, I and H, called, respectively, the index glass and the horizon glass, both of which are firmly fixed perpendicular to the plane of the instrument. The index glass is attached to the movable arm, ID, and turns as this is moved along the graduated limb, EF. This arm also carries a *vernier*, at D, a contrivance which, like the micrometer, enables us to take off minute parts of the spaces into which the limb is divided. The horizon glass, H, consists of two parts; the upper part being transparent or open, so that the eye, looking through the telescope, T, can see through it a distant body, as a star at S, while the lower part is a reflector.

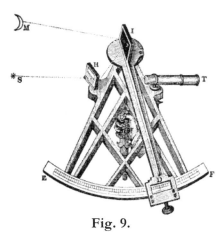

Fig. 9.

Suppose it were required to measure the angular distance between the moon and a certain star,—the moon being at M, and the star at S. The instrument is held firmly in the hand, so that the eye, looking through the telescope, sees the star, S, through the transparent part of the horizon glass. Then the movable arm, ID, is moved from F towards E, until the image of M is reflected down to S, when the number of degrees and parts of a degree reckoned on the limb, from F to the index at D, will show the angular distance between the two bodies.

LETTER VI.

TIME AND THE CALENDAR.

"From old Eternity's mysterious orb
Was Time cut off, and cast beneath the skies."—*Young.*

HAVING hitherto been conversant only with the many fine and sentimental things which the poets have sung respecting Old Time, perhaps you will find some difficulty in bringing down your mind to the calmer consideration of what time really is, and according to what different standards it is measured for different purposes. You will not, however, I think, find the subject even in our matter-of-fact and unpoetical way of treating it, altogether uninteresting. What, then, is time? *Time is a measured portion of indefinite duration.* It consists of equal portions cut off from eternity, as a line on the surface of the earth is separated from its contiguous portions that constitute a great circle of the sphere, by applying to it a two-foot scale; or as a few yards of cloth are measured off from a piece of unknown or indefinite extent.

Any thing, or any event which takes place at equal intervals, may become a measure of time. Thus, the pulsations of the wrist, the flowing of a given quantity of sand from one vessel to another, as in the hourglass, the beating of a pendulum, and the revolution of a star, have been severally employed as measures of time. But the great standard of time is the period of the revolution of the earth on its axis, which, by the most exact observations, is found to be always the same. I have anticipated a little of this subject, in giving an account of the transit instrument and clock, but I propose, in this letter, to enter into it more at large.

The time of the earth's revolution on its axis, as already explained, is called a sidereal day, and is determined by the revolution of a star in the heavens. This interval is divided into twenty-four *sidereal* hours. Observations taken on numerous stars, in different ages of the world, show that they all perform their diurnal revolution in the same time, and that their motion, during any part of the revolution, is always uniform. Here, then, we have an exact measure of time, probably more exact than any thing which can be devised by art. *Solar time* is reckoned by the apparent revolution of the sun from the meridian round to the meridian again. Were the sun stationary in the heavens, like a fixed star, the time of its apparent revolution would be equal to the revolution of the earth on its axis, and the solar and the sidereal days would be equal. But, since the sun passes from west to east, through three hundred and sixty degrees, in three hundred and sixty-five and one fourth days, it moves eastward nearly one degree a day. While, therefore, the earth is turning round on its axis, the sun is moving in the same direction, so that, when we have come round under the same celestial meridian from which we started, we do not find the sun there, but he has moved eastward nearly a degree, and the earth must perform so much more than one complete revolution, before we come under the sun again. Now, since we move, in the diurnal revolution, fifteen degrees in sixty minutes, we must pass over one degree in four minutes. It takes, therefore, four minutes for us to *catch up* with the sun, after we have made one complete revolution. Hence the solar day is about four minutes longer than the sidereal; and if we were to reckon the sidereal day twenty-four hours, we should reckon the solar day twenty-four hours four minutes. To suit the purposes of society at large, however, it is found more convenient to reckon the solar days twenty-four hours, and throw the fraction into the sidereal day. Then,

$$24h. \ 4m. : 24h. :: 24h. : 23h. \ 56m. \ 4s.$$

That is, when we reduce twenty-four hours and four minutes to twenty-four hours, the same proportion will require that we reduce the sidereal day from twenty-four hours to twenty-three hours fifty-six minutes four seconds; or, in other words, a sidereal day is such a part of a solar day. The solar days, however, do not always differ from the sidereal by precisely the same fraction, since they are not constantly of the same length. Time, as measured by the sun, is called *apparent time*, and a clock so regulated as always to keep exactly with the sun, is said to keep apparent time. *Mean time* is time reckoned by the *average* length of all the solar days throughout the year. This is the period which constitutes the *civil* day of twenty-four hours, beginning when the sun is on the lower meridian, namely, at twelve o'clock at night, and counted by twelve hours from the lower to the upper meridian, and from the upper to the lower. The *astronomical* day is the apparent solar day counted through the whole twenty-four hours, (instead of by periods of twelve hours each, as in the civil day,) and begins at noon. Thus it is now the tenth of June, at nine o'clock, A.M., according to civil time; but we have not yet reached the tenth of June by astronomical time, nor shall we, until noon to-day; consequently, it is now June ninth, twenty-first hour of astronomical time. Astronomers, since so many of their observations are taken on the meridian, are always supposed to look towards the south. Geographers, having formerly been conversant only with the northern hemisphere, are always understood to be looking towards the north. Hence, left and right, when applied to the astronomer, mean east and west, respectively; but to the geographer the right is east, and the left, west.

Clocks are usually regulated so as to indicate mean solar time; yet, as this is an artificial period not marked off, like the sidereal day, by any natural event, it is necessary to know how much is to be added to, or subtracted from, the apparent solar time, in order to give the corresponding mean time. The interval, by which apparent time differs from mean time, is called the *equation of time*. If one clock is so constructed as to keep exactly with the sun, going faster or slower, according as the lengths of the solar days vary, and another clock is regulated to mean time, then the difference of the two clocks, at any period, would be the equation of time for that moment. If the apparent clock were *faster* than the mean, then the equation of time must be subtracted; but if the apparent clock were slower than the mean, then the equation of time must be added, to give the mean time. The two clocks would differ most about the third of November, when the apparent time is sixteen and one fourth minutes greater than the mean. But since apparent time is sometimes greater and sometimes less than mean time, the two must obviously be sometimes equal to each other. This is, in fact, the case four times a year, namely, April fifteenth, June fifteenth, September first, and December twenty-fourth.

Astronomical clocks are made of the best workmanship, with every advantage that can promote their regularity. Although they are brought to an astonishing degree of accuracy, yet they are not as regular in their movements as the stars are, and their accuracy requires to be frequently tested. The transit instrument itself, when once accurately placed in the meridian, affords the means of testing the correctness of the clock, since one revolution of a star, from the meridian to the meridian again, ought to correspond exactly to twenty-four hours by the clock, and to continue the same, from day to day; and the right ascensions of various stars, as they cross the meridian, ought to be such by the clock, as they are given in the tables, where they are stated according to the accurate determinations of astronomers. Or, by taking the difference of any two stars, on successive days, it will be seen whether the going of the clock is uniform for that part of the day; and by taking the right ascensions of different pairs of stars, we may learn the rate of the clock at various parts of the day. We thus learn, not only whether the clock accurately measures the length of the sidereal day, but also whether it goes uniformly from hour to hour.

Although astronomical clocks have been brought to a great degree of perfection, so as hardly to vary a second for many months, yet none are absolutely perfect, and most are so far from it, as to require to be corrected by means of the transit instrument, every few days. Indeed, for the nicest observations, it is usual not to attempt to bring the clock to a state of absolute correctness, but, after bringing it as near to such a state as can conveniently be done, to ascertain how much it gains or loses in a day; that is, to ascertain the *rate* of its going, and to make allowance accordingly.

Having considered the manner in which the smaller divisions of time are measured, let us now take a hasty glance at the larger periods which compose the calendar.

As a *day* is the period of the revolution of the earth on its axis, so a *year* is the period of the revolution of the earth around the sun. This time, which constitutes the *astronomical year*, has been ascertained with great exactness, and found to be three hundred and sixty-five days five hours forty-eight minutes and fifty-one seconds. The most ancient nations determined the number of days in the year by means of the *stylus*, a perpendicular rod which casts its shadow on a smooth plane bearing a meridian line. The time when the shadow was shortest, would indicate the day of the Summer solstice; and the number of days which elapsed, until the shadow returned to the same length again, would show the number of days in the year. This was found to be three hundred and sixty-five whole days, and accordingly, this period was adopted for the civil year. Such a difference, however, between the civil and astronomical years, at length threw all dates into confusion. For if, at first, the Summer solstice happened on the twenty-first of June, at the end of four years, the sun would not have reached the solstice until the twenty-second of June; that is, it would have been behind its time. At the end of the next four years, the solstice would fall on the twenty-third; and in process of time, it would fall successively on every day of the year. The same would be true of any other fixed date.

Julius Cæsar, who was distinguished alike for the variety and extent of his knowledge, and his skill in arms, first attempted to make the calendar conform to the motions of the sun.

"Amidst the hurry of tumultuous war,The stars, the gods, the heavens, were still his care."

Aided by Sosigenes, an Egyptian astronomer, he made the first correction of the calendar, by introducing an additional day every fourth year, making February to consist of twenty-nine instead of twenty-eight days, and of course the whole year to consist of three hundred and sixty-six days. This fourth year was denominated *Bissextile*, because the sixth day before the Kalends of March was reckoned twice. It is also called Leap Year.

The Julian year was introduced into all the civilized nations that submitted to the Roman power, and continued in general use until the year 1582. But the true correction was not six hours, but five hours forty-nine minutes; hence the addition was too great by eleven minutes. This small fraction would amount in one hundred years to three fourths of a day, and in one thousand years to more than seven days. From the year 325 to the year 1582, it had, in fact, amounted to more than ten days; for it was known that, in 325, the vernal equinox fell on the twenty-first of March, whereas, in 1582, it fell on the eleventh. It was ordered by the Council of Nice, a celebrated ecclesiastical council, held in the year 325, that Easter should be celebrated upon the first Sunday after the first full moon, next following the vernal equinox; and as certain other festivals of the Romish Church were appointed at particular seasons of the year, confusion would result from such a want of constancy between any fixed date and a particular season of the year. Suppose, for example, a festival accompanied by numerous religious ceremonies, was decreed by the Church to be held at the time when the sun crossed the equator in the Spring, (an event hailed with great joy, as the harbinger of the return of Summer,) and that, in the year 325, March twenty-first was designated as the time for holding the festival, since, at that period, it was on the twenty-first of March when the sun reached the equinox; the next year, the sun would reach the equinox a little sooner than the twenty-first of March, only eleven minutes, indeed, but still amounting in twelve hundred years to ten days; that is, in 1582, the sun reached the equinox on the eleventh of March. If, therefore, they should continue to observe the twenty-first as a religious festival in honor of this event, they would commit the absurdity of celebrating it ten days after it had passed by. Pope Gregory the Thirteenth, who was then at the head of the Roman See, was a man of science, and undertook to reform the calendar, so that fixed dates would always correspond to the same seasons of the year. He first decreed, that the year should be brought forward ten days, by reckoning the fifth of October the fifteenth; and, in order to prevent the calendar from falling into confusion afterwards, he prescribed the following rule: *Every year whose number is not divisible by four, without a remainder, consists of three hundred and sixty-five days; every year which is so divisible, but is not divisible by one hundred, of three hundred and sixty-six; every year divisible by one hundred, but not by four hundred, again, of three hundred and sixty-five; and every year divisible by four hundred, of three hundred and sixty-six.*

Thus the year 1838, not being divisible by four, contains three hundred and sixty-five days, while 1836 and 1840 are leap years. Yet, to make every fourth year consist of three hundred and sixty-six days would increase it too much, by about three fourths of a day in a century; therefore every hundredth year has only three hundred and sixty-five days. Thus 1800, although divisible by four, was not a leap year, but a common year. But we have allowed a *whole* day in a hundred years, whereas we ought to have allowed only *three fourths* of a day. Hence, in four hundred years, we should allow a day too much, and therefore, we let the four hundredth remain a leap year. This rule involves an error of less than a day in four thousand two hundred and thirty-seven years.

The Pope, who, you will recollect, at that age assumed authority over all secular princes, issued his decree to the reigning sovereigns of Christendom, commanding the observance of the calendar as reformed by him. The decree met with great opposition among the Protestant States, as they recognised in it a new exercise of ecclesiastical tyranny; and some of them, when they received it, made it expressly understood, that their acquiescence should not be construed as a submission to the Papal authority.

In 1752, the Gregorian year, or *New Style*, was established in Great Britain by act of Parliament; and the dates of all deeds, and other legal papers, were to be made according to it. As above a century had then passed since the first introduction of the new style, eleven days were suppressed, the third of September being called the fourteenth. By the same act, the beginning of the year was changed from March twenty-fifth to January first. A few persons born previously to 1752 have come down to our day, and we frequently see inscriptions on tombstones of those whose time of birth is recorded in old style. In order to make this correspond to our present mode of reckoning, we must add eleven days to the date. Thus the same event would be June twelfth of old style, or June twenty-third of new style; and if an event occurred between January first and March twenty-fifth, the date of the year would be advanced one, since February 1st, 1740, O.S. would be February 1st, 1741, N.S. Thus, General Washington was born February 11th, 1731, O.S., or February 22d, 1732, N.S. If we inquire how any present event may be made to correspond in date to the old style, we must subtract twelve days, and put the year back one, if the event lies between January first and March twenty-fifth. Thus, June tenth, N.S. corresponds to May twenty-ninth, O.S.; and March 20th, 1840, to March 8th, 1839. France, being a Roman Catholic country, adopted the new style soon after it was decreed by the Pope; but Protestant countries, as we have seen, were much slower in adopting it; and Russia, and the Greek Church generally, still adhere to the old style. In order, therefore, to make the Russian dates correspond to ours, we must add to them twelve days.

It may seem to you very remarkable, that so much pains should have been bestowed upon this subject; but without a correct and uniform standard of time, the dates of deeds, commissions, and all legal papers; of fasts and festivals, appointed by ecclesiastical authority; the returns of seasons, and the records of history,—must all fall into inextricable confusion. To change the observance of certain religious feasts, which have been long fixed to particular days, is looked upon as an impious innovation; and though the times of the events, upon which these ceremonies depend, are utterly unknown, it is still insisted upon by certain classes in England, that the Glastenbury thorn blooms on Christmas day.

Although the ancient Grecian calendar was extremely defective, yet the common people were entirely averse to its reformation. Their superstitious adherence to these errors was satirized by Aristophanes, in his comedy of the Clouds. An actor, who had just come from Athens, recounts that he met with Diana, or the moon, and found her extremely incensed, that they did not regulate her course better. She complained, that the order of Nature was changed, and every thing turned topsyturvy. The gods no longer knew what belonged to them; but, after paying their visits on certain feast-days, and expecting to meet with good cheer, as usual, they were under the disagreeable necessity of returning back to heaven without their suppers.

Among the Greeks, and other ancient nations, the length of the year was generally regulated by the course of the moon. This planet, on account of the different appearances which she exhibits at her full, change, and quarters, was considered by them as best adapted of any of the celestial bodies for this purpose. As one lunation, or revolution of the moon around the earth, was found to be completed in about twenty-nine and one half days, and twelve of these periods being supposed equal to one revolution of the sun, their months were made to consist of twenty-nine and thirty days alternately, and their year of three hundred and fifty-four days. But this disagreed with the annual revolution of the sun, which must evidently govern the seasons of the year, more than eleven days. The irregularities, which such a mode of reckoning would occasion, must have been too obvious not to have been observed. For, supposing it to have been settled, at any particular time, that the beginning of the year should be in the Spring; in about sixteen years afterwards, the beginning would have been in Autumn; and in thirty-three or thirty-four years, it would have gone backwards through all the seasons, to Spring again. This defect they attempted to rectify, by introducing a number of days, at certain times, into the calendar, as occasion required, and putting the beginning of the year forwards, in order to make it agree with the course of the sun. But as these additions, or *intercalations*, as they were called, were generally consigned to the care of the priests, who, from motives of interest or superstition, frequently omitted them, the year was made long or short at pleasure.

The *week* is another division of time, of the highest antiquity, which, in almost all countries, has been made to consist of seven days; a period supposed by some to have been traditionally derived from the creation of the world; while others imagine it to have been regulated by the phases of the moon. The names, Saturday, Sunday, and Monday, are obviously derived from Saturn, the Sun, and the Moon; while Tuesday, Wednesday, Thursday, and Friday, are the days of Tuisco, Woden, Thor, and Friga, which are Saxon names for Mars, Mercury, Jupiter, and Venus.[4]

The common year begins and ends on the same day of the week; but leap year ends one day later than it began. Fifty-two weeks contain three hundred and sixty-four days; if, therefore, the year begins on Tuesday, for example, we should complete fifty-two weeks on Monday, leaving one day, (Tuesday,) to complete the year, and the following year would begin on Wednesday. Hence, any day of the month is one day later in the week, than the corresponding day of the preceding year. Thus, if the sixteenth of November, 1838, falls on Friday, the sixteenth of November, 1837, fell on Thursday, and will fall, in 1839, on Saturday. But if leap year begins on Sunday, it ends on Monday, and the following year begins on Tuesday; while any given day of the month is two days later in the week than the corresponding date of the preceding year.

LETTER VII.

FIGURE OF THE EARTH.

"He took the golden compasses, prepared
In God's eternal store, to circumscribe
This universe, and all created things;
One foot he centred, and the other turned
Round through the vast profundity obscure,

IN the earliest ages, the earth was regarded as one continued plane; but, at a comparatively remote period, as five hundred years before the Christian era, astronomers began to entertain the opinion that the earth is round. We are able now to adduce various arguments which severally prove this truth. First, when a ship is coming in from sea, we first observe only the very highest parts of the ship, while the lower portions come successively into view. Were the earth a continued plane, the lower parts of the ship would be visible as soon as the higher, as is evident from Fig. 10, page 70.

Fig. 10.

Since light comes to the eye in straight lines, by which objects become visible, it is evident, that no reason exists why the parts of the ship near the water should not be seen as soon as the upper parts. But if the earth be a sphere, then the line of sight would pass above the deck of the ship, as is represented in Fig. 11; and as the ship drew nearer to land, the lower parts would successively rise above this line and come into view exactly in the manner known to observation. Secondly, in a lunar eclipse, which is occasioned by the moon's passing through the earth's shadow, the figure of the shadow is seen to be spherical, which could not be the case unless the earth itself were round. Thirdly, navigators, by steering continually in one direction, as east or west, have in fact come round to the point from which they started, and thus confirmed the fact of the earth's rotundity beyond all question. One may also reach a given place on the earth, by taking directly opposite courses. Thus, he may reach Canton in China, by a westerly route around Cape Horn, or by an easterly route around the Cape of Good Hope. All these arguments severally prove that the earth is round.

Fig. 11.

But I propose, in this Letter, to give you some account of the unwearied labors which have been performed to ascertain the *exact* figure of the earth; for although the earth is properly described in general language as round, yet it is not an exact sphere. Were it so, all its diameters would be equal; but it is known that a diameter drawn through the equator exceeds one drawn from pole to pole, giving to the earth the form of a *spheroid*,—a figure resembling an orange, where the ends are flattened a little and the central parts are swelled out.

Although it would be a matter of very rational curiosity, to investigate the precise shape of the planet on which Heaven has fixed our abode, yet the immense pains which has been bestowed on this subject has not all arisen from mere curiosity. No accurate measurements can be taken of the distances and magnitudes of the heavenly bodies, nor any exact determinations made of their motions, without a knowledge of the exact figure of the earth; and hence is derived a powerful motive for ascertaining this element with all possible precision.

The first satisfactory evidence that was obtained of the exact figure of the earth was derived from reasoning on the effects of the earth's *centrifugal force*, occasioned by its rapid revolution on its own axis. When water is whirled in a pail, we see it recede from the centre and accumulate upon the sides of the vessel; and when a millstone is whirled rapidly, since the portions of the stone furthest from the centre revolve much more rapidly than those near to it, their greater tendency to recede sometimes makes them fly off with a violent explosion. A case, which comes still nearer to that of the earth, is exhibited by a mass of clay revolving on a potter's wheel, as seen in the process of making earthen vessels. The mass swells out in the middle, in consequence of the centrifugal force exerted upon it by a rapid motion. Now, in the diurnal revolution, the equatorial parts of the earth move at the rate of about one thousand miles per hour, while the poles do not move at all; and since, as we take points at successive distances from the equator towards the pole, the rate at which these points move grows constantly less and less; and since, in revolving bodies, the centrifugal force is proportioned to the velocity, consequently, those parts which move with the greatest rapidity will be more affected by this force than those which move more slowly. Hence, the equatorial regions must be higher from the centre than the polar regions; for, were not this the case, the waters on the surface of the earth would be thrown towards the equator, and be piled up there, just as water is accumulated on the sides of a pail when made to revolve rapidly.

Huyghens, an eminent astronomer of Holland, who investigated the laws of centrifugal forces, was the first to infer that such must be the actual shape of the earth; but to Sir Isaac Newton we owe the full developement of this doctrine. By combining the reasoning derived from the known laws of the centrifugal force with arguments derived from the principles of universal gravitation, he concluded that the distance through the earth, in the direction of the equator, is greater than that in the direction of the poles. He estimated the difference to be about thirty-four miles.

But it was soon afterwards determined by the astronomers of France, to ascertain the figure of the earth by actual measurements, specially instituted for that purpose. Let us see how this could be effected. If we set out at the equator and travel towards the pole, it is easy to see when we have advanced one degree of latitude, for this will be indicated by the rising of the north star, which appears in the horizon when the spectator stands on the equator, but rises in the same proportion as he recedes from the equator, until, on reaching the pole, the north star would be seen directly over head. Now, were the earth a perfect sphere, the meridian of the earth would be a perfect circle, and the distance between any two places, differing one degree in latitude, would be exactly equal to the distance between any other two places, differing in latitude to the same amount. But if the earth be a spheroid, flattened at the poles, then a line encompassing the earth from north to south, constituting the terrestrial meridian, would not be a perfect circle, but an ellipse or oval, having its longer diameter through the equator, and its shorter through the poles. The part of this curve included between two radii, drawn from the centre of the earth to the celestial meridian, at angles one degree asunder, would be greater in the polar than in the equatorial region; that is, the degrees of the meridian would lengthen towards the poles.

The French astronomers, therefore, undertook to ascertain by actual measurements of arcs of the meridian, in different latitudes, whether the degrees of the meridian are of uniform length, or, if not, in what manner they differ from each other. After several indecisive measurements of an arc of the meridian in France, it was determined to effect simultaneous measurements of arcs of the meridian near the equator, and as near as possible to the north pole, presuming that if degrees of the meridian, in different latitudes, are really of different lengths, they will differ most in points most distant from each other. Accordingly, in 1735, the French Academy, aided by the government, sent out two expeditions, one to Peru and the other to Lapland. Three distinguished mathematicians, Bouguer, La Condamine, and Godin, were despatched to the former place, and four others, Maupertius, Camus, Clairault, and Lemonier, were sent to the part of Swedish Lapland which lies at the head of the Gulf of Tornea, the northern arm of the Baltic. This commission completed its operations several years sooner than the other, which met with greater difficulties in the way of their enterprise. Still, the northern detachment had great obstacles to contend with, arising particularly from the extreme length and severity of their Winters. The measurements, however, were conducted with care and skill, and the result, when compared with that obtained for the length of a degree in France, plainly indicated, by its greater amount, a compression of the earth towards the poles.

Mean-while, Bouguer and his party were prosecuting a similar work in Peru, under extraordinary difficulties. These were caused, partly by the localities, and partly by the ill-will and indolence of the inhabitants. The place selected for their operations was in an elevated valley between two principal chains of the Andes. The lowest point of their arc was at an elevation of a mile and a half above the level of the sea; and, in some instances, the heights of two neighboring signals differed more than a mile. Encamped upon lofty mountains, they had to struggle against storms, cold, and privations of every description, while the invincible indifference of the Indians, they were forced to employ, was not to be shaken by the fear of punishment or the hope of reward. Yet, by patience and ingenuity, they overcame all obstacles, and executed with great accuracy one of the most important operations, of this nature, ever undertaken. To accomplish this, however, took them nine years; of which, three were occupied in determining the latitudes alone.[5]

I have recited the foregoing facts, in order to give you some idea of the unwearied pains which astronomers have taken to ascertain the exact figure of the earth. You will find, indeed, that all their labors are characterized by the same love of accuracy. Years of toilsome watchings, and incredible labor of computation, have been undergone, for the sake of arriving only a few seconds nearer to the truth.

The length of a degree of the meridian, as measured in Peru, was less than that before determined in France, and of course less than that of Lapland; so that the spheroidal figure of the earth appeared now to be ascertained by actual measurement. Still, these measures were too few in number, and covered too small a portion of the whole quadrant from the equator to the pole, to enable astronomers to ascertain the exact law of curvature of the meridian, and therefore similar measurements have since been prosecuted with great zeal by different nations, particularly by the French and English. In 1764, two English mathematicians of great eminence, Mason and Dixon, undertook the measurement of an arc in Pennsylvania, extending more than one hundred miles.

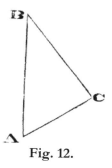

Fig. 12.

These operations are carried on by what is called a system of *triangulation*. Without some knowledge of trigonometry, you will not be able fully to understand this process; but, as it is in its nature somewhat curious, and is applied to various other geographical measurements, as well as to the determination of arcs of the meridian, I am desirous that you should understand its general principles. Let us reflect, then, that it must be a matter of the greatest difficulty, to execute with exactness the measurement of a line of any great length in one continued direction on the earth's surface. Even if we select a level and open country, more or less inequalities of surface will occur; rivers must be crossed, morasses must be traversed, thickets must be penetrated, and innumerable other obstacles must be surmounted; and finally, every time we apply an artificial measure, as a rod, for example, we obtain a result not absolutely perfect. Each error may indeed be very small, but small errors, often repeated, may produce a formidable aggregate. Now, one unacquainted with trigonometry can easily understand the fact, that, when we know certain parts of a triangle, we can find the other parts by calculation; as, in the rule of three in arithmetic, we can obtain the fourth term of a proportion, from having the first three terms given. Thus, in the triangle A B C, Fig. 12, if we know the side A B, and the angles at A and B, we can find by computation, the other sides, A C and B C, and the remaining angle at C. Suppose, then, that in measuring an arc of the meridian through any country, the line were to pass directly through A B, but the ground was so obstructed between A and B, that we could not possibly carry our measurement through it. We might then measure another line, as A C, which was accessible, and with a compass take the bearing of B from the points A and C, by which means we should learn the value of the angles at A and C. From these data we might calculate, by the rules of trigonometry, the exact length of the line A B. Perhaps the ground might be so situated, that we could not reach the point B, by any route; still, if it could be seen from A and C, it would be all we should want. Thus, in conducting a trigonometrical survey of any country, conspicuous signals are placed on elevated points, and the bearings of these are taken from the extremities of a known line, called the base, and thus the relative situation of various places is accurately determined. Were we to undertake to run an exact north and south line through any country, as New England, we should select, near one extremity, a spot of ground favorable for actual measurement, as a level, unobstructed plain; we should provide a measure whose length in feet and inches was determined with the greatest possible precision, and should apply it with the utmost care. We should thus obtain a *base line*. From the extremities of this line, we should take (with some appropriate instrument) the bearing of some signal at a greater or less distance, and thus we should obtain one side and two angles of a triangle, from which we could find, by the rules of trigonometry, either of the unknown sides. Taking this as a new base, we might take the bearing of another signal, still further on our way, and thus proceed to run the required north and south line, without actually measuring any thing more than the first, or base line.

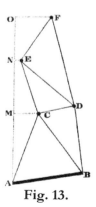

Fig. 13.

Thus, in Fig. 13, we wish to measure the distance between the two points A and O, which are both on the same meridian, as is known by their having the same longitude; but, on account of various obstacles, it would be found very inconvenient to measure this line directly, with a rod or chain, and even if we could do it, we could not by this method obtain nearly so accurate a result, as we could by a series of triangles, where, after the base line was measured, we should have nothing else to measure except angles, which can be determined, by observation, to a greater degree of exactness, than lines. We therefore, in the first place, measure the base line, A B, with the utmost precision. Then, taking the bearing of some signal at C from A and B, we obtain the means of calculating the side B C, as has been already explained. Taking B C as a new base, we proceed, in like manner, to determine successively the sides C D, D E, and E F, and also A C, and C E. Although A C is not in the direction of the meridian, but considerably to the east of it, yet it is easy to find the corresponding distance on the meridian, A M; and in the same manner we can find the portions of the meridian M N and N O, corresponding respectively to C E and E F. Adding these several parts of the meridian together, we obtain the length of the arc from A to O, in miles; and by observations on the north star, at each extremity of the arc, namely, at A and at O, we could determine the difference of latitude between these two points. Suppose, for example, that the distance between A and O is exactly five degrees, and that the length of the intervening line is three hundred and forty-seven miles; then, dividing the latter by the former number, we find the length of a degree to be sixty-nine miles and four tenths. To take, however, a few of the results actually obtained, they are as follows:

Places of observation.	Latitude.	Length of a deg. in miles.
Peru,	00° 00' 00"	68.732
Pennsylvania,	39 12 00	68.896
France,	46 12 00	69.054
England,	51 29 54½	69.146
Sweden,	66 20 10	69.292

This comparison shows, that the length of a degree gradually increases, as we proceed from the equator towards the pole. Combining the results of various estimates, the dimensions of the terrestrial spheroid are found to be as follows:

Equatorial diameter,	7925.648 miles.
Polar diameter,	7899.170 "
Average diameter,	7912.409 "

The difference between the greatest and the least is about twenty-six and one half miles, which is about one two hundred and ninety-ninth part of the greatest. This fraction is denominated the *ellipticity* of the earth,—being the excess of the equatorial over the polar diameter.

The operations, undertaken for the purpose of determining the figure of the earth, have been conducted with the most refined exactness. At any stage of the process, the length of the last side, as obtained by calculation, may be actually measured in the same manner as the base from which the series of triangles commenced. When thus measured, it is called the *base of verification*. In some surveys, the base of verification, when taken at a distance of four hundred miles from the starting point, has not differed more than one foot from the same line, as determined by calculation.

Another method of arriving at the exact figure of the earth is, by observations with the *pendulum*. If a pendulum, like that of a clock, be suspended, and the number of its vibrations per hour be counted, they will be found to be different in different latitudes. A pendulum that vibrates thirty-six hundred times per hour, at the equator, will vibrate thirty-six hundred and five and two thirds times, at London, and a still greater number of times nearer the north pole. Now, the vibrations of the pendulum are produced by the force of gravity. Hence their comparative number at different places is a measure of the relative forces of gravity at those places. But when we know the relative forces of gravity at different places, we know their relative distances from the centre of the earth; because the nearer a place is to the centre of the earth, the greater is the force of gravity. Suppose, for example, we should count the number of vibrations of a pendulum at the equator, and then carry it to the north pole, and count the number of vibrations made there in the same time,—we should be able, from these two observations, to estimate the relative forces of gravity at these two points; and, having the relative forces of gravity, we can thence deduce their relative distances from the centre of the earth, and thus obtain the polar and equatorial diameters. Observations of this kind have been taken with the greatest accuracy, in many places on the surface of the earth, at various distances from each other, and they lead to the same conclusions respecting the figure of the earth, as those derived from measuring arcs of the meridian. It is pleasing thus to see a great truth, and one apparently beyond the pale of human investigation, reached by two routes entirely independent of each other. Nor, indeed, are these the only proofs which have been discovered of the spheroidal figure of the earth. In consequence of the accumulation of matter above the equatorial regions of the earth, a body weighs less there than towards the poles, being further removed from the centre of the earth. The same accumulation of matter, by the force of attraction which it exerts, causes slight inequalities in the motions of the moon; and since the amount of these becomes a measure of the force which produces them, astronomers are able, from these inequalities, to calculate the exact quantity of the matter thus accumulated, and hence to determine the figure of the earth. The result is not essentially different from that obtained by the other methods. Finally, the shape of the earth's shadow is altered, by its spheroidal figure,—a circumstance which affects the time and duration of a lunar eclipse. All these different and independent phenomena afford a pleasing example of the harmony of truth. The known effects of the centrifugal force upon a body revolving on its axis, like the earth, lead us to infer that the earth is of a spheroidal figure; but if this be the fact, the pendulum ought to vibrate faster near the pole than at the equator, because it would there be nearer the centre of the earth. On trial, such is found to be the case. If, again, there be such an accumulation of matter about the equatorial regions, its effects ought to be visible in the motions of the moon, which it would influence by its gravity; and there, also, its effects are traced. At length, we apply our measures to the surface of the earth itself, and find the same fact, which had thus been searched out among the hidden things of Nature, here palpably exhibited before our eyes. Finally, on estimating from these different sources, what the exact amount of the compression at the poles must be, all bring out nearly one and the same result. This truth, so harmonious in itself, takes along with it, and establishes, a thousand other truths on which it rests.

LETTER VIII.

DIURNAL REVOLUTIONS.

"To some she taught the fabric of the sphere,
The changeful moon, the circuit of the stars,
The golden zones of heaven."—*Akenside.*

WITH the elementary knowledge already acquired, you will now be able to enter with pleasure and profit on the various interesting phenomena dependent on the revolution of the earth on its axis and around the sun. The apparent diurnal revolution of the heavenly bodies, from east to west, is owing to the actual revolution of the earth on its own axis, from west to east. If we conceive of a radius of the earth's equator extended until it meets the concave sphere of the heavens, then, as the earth revolves, the extremity of this line would trace out a curve on the face of the sky; namely, the celestial equator. In curves parallel to this, called the *circles of diurnal revolution*, the heavenly bodies actually *appear* to move, every star having its own peculiar circle. After you have first rendered familiar the real motion of the earth from west to east, you may then, without danger of misapprehension, adopt the common language, that all the heavenly bodies revolve around the earth once a day, from east to west, in circles parallel to the equator and to each other.

I must remind you, that the time occupied by a star, in passing from any point in the meridian until it comes round to the same point again, is called a *sidereal day*, and measures the period of the earth's revolution on its axis. If we watch the returns of the same star from day to day, we shall find the intervals exactly equal to each other; that is, *the sidereal days are all equal.* Whatever star we select for the observation, the same result will be obtained. The stars, therefore, always keep the same relative position, and have a common movement round the earth,—a consequence that naturally flows from the hypothesis that their *apparent* motion is all produced by a single *real* motion; namely, that of the earth. The sun, moon, and planets, as well as the fixed stars, revolve in like manner; but their returns to the meridian are not, like those of the fixed stars, at exactly equal intervals.

The *appearances* of the diurnal motions of the heavenly bodies are different in different parts of the earth,—since every place has its own horizon, and different horizons are variously inclined to each other. Nothing in astronomy is more apt to mislead us, than the obstinate habit of considering the horizon as a fixed and immutable plane, and of referring every thing to it. We should contemplate the earth as a huge globe, occupying a small portion of space, and encircled on all sides, at an immense distance, by the starry sphere. We should free our minds from their habitual proneness to consider one part of space as naturally *up* and another *down*, and view ourselves as subject to a force (gravity) which binds us to the earth as truly as though we were fastened to it by some invisible cords or wires, as the needle attaches itself to all sides of a spherical loadstone. We should dwell on this point, until it appears to us as truly up, in the direction B B, C C, D D, when one is at B, C, D, respectively, as in the direction A A, when he is at A, Fig. 14.

Let us now suppose the spectator viewing the diurnal revolutions from several different positions on the earth. On the *equator*, his horizon would pass through both poles; for the horizon cuts the celestial vault at ninety degrees in every direction from the zenith of the spectator; but the pole is likewise ninety degrees from his zenith, when he stands on the equator; and consequently, the pole must be in the horizon. Here, also, the celestial equator would coincide with the prime vertical, being a great circle passing through the east and west points. Since all the diurnal circles are parallel to the equator, consequently, they would all, like the equator be perpendicular to the horizon. Such a view of the heavenly bodies is called a right sphere, which may be thus defined: *a right sphere is one in which all the daily revolutions of the stars are in circles perpendicular to the horizon.*

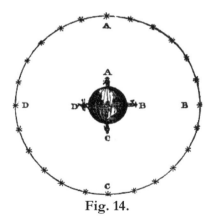

Fig. 14.

A right sphere is seen only at the equator. Any star situated in the celestial equator would appear to rise directly in the east, at midnight to be in the zenith of the spectator, and to set directly in the west. In proportion as stars are at a greater distance from the equator towards the pole, they describe smaller and smaller circles, until, near the pole, their motion is hardly perceptible.

If the spectator advances one degree from the equator towards the north pole, his horizon reaches one degree beyond the pole of the earth, and cuts the starry sphere one degree below the pole of the heavens, or below the north star, if that be taken as the place of the pole. As he moves onward towards the pole, his horizon continually reaches further and further beyond it, until, when he comes to the pole of the earth, and under the pole of the heavens, his horizon reaches on all sides to the equator, and coincides with it. Moreover, since all the circles of daily motion are parallel to the equator, they become, to the spectator at the pole, parallel to the horizon. Or, *a parallel sphere is that in which all the circles of daily motion are parallel to the horizon.*

To render this view of the heavens familiar, I would advise you to follow round in mind a number of separate stars, in their diurnal revolution, one near the horizon, one a few degrees above it, and a third near the zenith. To one who stood upon the north pole, the stars of the northern hemisphere would all be perpetually in view when not obscured by clouds, or lost in the sun's light, and none of those of the southern hemisphere would ever be seen. The sun would be constantly above the horizon for six months in the year, and the remaining six continually out of sight. That is, at the pole, the days and nights are each six months long. The appearances at the south pole are similar to those at the north.

A perfect parallel sphere can never be seen, except at one of the poles,—a point which has never been actually reached by man; yet the British discovery ships penetrated within a few degrees of the north pole, and of course enjoyed the view of a sphere nearly parallel.

As the circles of daily motion are parallel to the horizon of the pole, and perpendicular to that of the equator, so at all places between the two, the diurnal motions are oblique to the horizon. This aspect of the heavens constitutes an oblique sphere, which is thus defined: *an oblique sphere is that in which the circles of daily motion are oblique to the horizon.*

Suppose, for example, that the spectator is at the latitude of fifty degrees. His horizon reaches fifty degrees beyond the pole of the earth, and gives the same apparent elevation to the pole of the heavens. It cuts the equator and all the circles of daily motion, at an angle of forty degrees,—being always equal to what the altitude of the pole lacks of ninety degrees: that is, it is always equal to the co-altitude of the pole. Thus, let H O, Fig. 15, represent the horizon, E Q the equator, and P P the axis of the earth. Also, *l l*, *m m*, *n n*, parallels of latitude. Then the horizon of a spectator at Z, in latitude fifty degrees, reaches to fifty degrees beyond the pole; and the angle E C H, which the equator makes with the horizon, is forty degrees,—the complement of the latitude. As we advance still further north, the elevation of the diurnal circle above the horizon grows less and less, and consequently, the motions of the heavenly bodies more and more oblique to the horizon, until finally, at the pole, where the latitude is ninety degrees, the angle of elevation of the equator vanishes, and the horizon and the equator coincide with each other, as before stated.

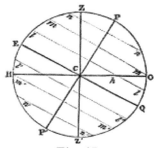

Fig. 15.

The circle of perpetual apparition is the boundary of that space around the elevated pole, where the stars never set. Its distance from the pole is equal to the latitude of the place. For, since the altitude of the pole is equal to the latitude, a star, whose polar distance is just equal to the latitude, will, when at its lowest point, only just reach the horizon; and all the stars nearer the pole than this will evidently not descend so far as the horizon. Thus *m m*, Fig. 15, is the circle of perpetual apparition, between which and the north pole, the stars never set, and its distance from the pole, O P, is evidently equal to the elevation of the pole, and of course to the latitude.

In the opposite hemisphere, a similar part of the sphere adjacent to the depressed pole never rises. Hence, *the circle of perpetual occultation is the boundary of that space around the depressed pole, within which the stars never rise.*

Thus *m´ m´*, Fig. 15, is the circle of perpetual occultation, between which and the south pole, the stars never rise.

In an oblique sphere, the horizon cuts the circles of daily motion unequally. Towards the elevated pole, more than half the circle is above the horizon, and a greater and greater portion, as the distance from the equator is increased, until finally, within the circle of perpetual apparition, the whole circle is above the horizon. Just the opposite takes place in the hemisphere next the depressed pole. Accordingly, when the sun is in the equator, as the equator and horizon, like all other great circles of the sphere, bisect each other, the days and nights are equal all over the globe. But when the sun is north of the equator, the days become longer than the nights, but shorter, when the sun is south of the equator. Moreover, the higher the latitude, the greater is the inequality in the lengths of the days and nights. By examining Fig. 15, you will easily see how each of these cases must hold good.

Most of the appearances of the diurnal revolution can be explained, either on the supposition that the celestial sphere actually turns around the earth once in twenty-four hours, or that this motion of the heavens is merely apparent, arising from the revolution of the earth on its axis, in the opposite direction,—a motion of which we are insensible, as we sometimes lose the consciousness of our own motion in a ship or steam-boat, and observe all external objects to be receding from us, with a common motion. Proofs, entirely conclusive and satisfactory, establish the fact, that it is the earth, and not the celestial sphere, that turns; but these proofs are drawn from various sources, and one is not prepared to appreciate their value, or even to understand some of them, until he has made considerable proficiency in the study of astronomy, and become familiar with a great variety of astronomical phenomena. To such a period we will therefore postpone the discussion of the earth's rotation on its axis.

While we retain the same place on the earth, the diurnal revolution occasions no change in our horizon, but our horizon goes round, as well as ourselves. Let us first take our station on the equator, at sunrise; our horizon now passes through both the poles and through the sun, which we are to conceive of as at a great distance from the earth, and therefore as cut, not by the terrestrial, but by the celestial, horizon. As the earth turns, the horizon dips more and more below the sun, at the rate of fifteen degrees for every hour; and, as in the case of the polar star, the sun appears to rise at the same rate. In six hours, therefore, it is depressed ninety degrees below the sun, bringing us directly under the sun, which, for our present purpose, we may consider as having all the while maintained the same fixed position in space. The earth continues to turn, and in six hours more, it completely reverses the position of our horizon, so that the western part of the horizon, which at sunrise was diametrically opposite to the sun, now cuts the sun, and soon afterwards it rises above the level of the sun, and the sun sets. During the next twelve hours, the sun continues on the invisible side of the sphere, until the horizon returns to the position from which it set out, and a new day begins.

Let us next contemplate the similar phenomena at the *poles*. Here the horizon, coinciding, as it does, with the equator, would cut the sun through its centre and the sun would appear to revolve along the surface of the sea, one half above and the other half below the horizon. This supposes the sun in its annual revolution to be at one of the equinoxes. When the sun is north of the equator, it revolves continually round in a circle, which, during a single revolution, appears parallel to the equator, and it is constantly day; and when the sun is south of the equator, it is, for the same reason, continual night.

When we have gained a clear idea of the appearances of the diurnal revolutions, as exhibited to a spectator at the equator and at the pole, that is, in a right and in a parallel sphere, there will be little difficulty in imagining how they must be in the intermediate latitudes, which have an oblique sphere.

The appearances of the sun and stars, presented to the inhabitants of different countries, are such as correspond to the sphere in which they live. Thus, in the fervid climates of India, Africa, and South America, the sun mounts up to the highest regions of the heavens, and descends directly downwards, suddenly plunging beneath the horizon. His rays, darting almost vertically upon the heads of the inhabitants, strike with a force unknown to the people of the colder climates; while in places remote from the equator, as in the north of Europe, the sun, in Summer, rises very far in the north, takes a long circuit towards the south, and sets as far northward in the west as the point where it rose on the other side of the meridian. As we go still further north, to the northern parts of Norway and Sweden, for example, to the confines of the frigid zone, the Summer's sun just grazes the northern horizon, and at noon appears only twenty-three and one half degrees above the southern. On the other hand, in mid-winter, in the north of Europe, as at St. Petersburgh, the day dwindles almost to nothing,— lasting only while the sun describes a very short arc in the extreme south. In some parts of Siberia and Iceland, the only day consists of a little glimmering of the sun on the verge of the southern horizon, at noon.

LETTER IX.

PARALLAX AND REFRACTION.

"Go, wondrous creature! mount where science guides,
Go measure earth, weigh air, and state the tides;
Instruct the planets in what orbs to run,
Correct old Time, and regulate the sun."—*Pope.*

I THINK you must have felt some astonishment, that astronomers are able to calculate the exact distances and magnitudes of the sun, moon, and planets. We should, at the first thought, imagine that such knowledge as this must be beyond the reach of the human faculties, and we might be inclined to suspect that astronomers practise some deception in this matter, for the purpose of exciting the admiration of the unlearned. I will therefore, in the present Letter, endeavor to give you some clear and correct views respecting the manner in which astronomers acquire this knowledge.

In our childhood, we all probably adopt the notion that the sky is a real dome of definite surface, in which the heavenly bodies are fixed. When any objects are beyond a certain distance from the eye, we lose all power of distinguishing, by our sight alone, between different distances, and cannot tell whether a given object is one million or a thousand millions of miles off. Although the bodies seen in the sky are in fact at distances extremely various,—some, as the clouds, only a few miles off; others, as the moon, but a few thousand miles; and others, as the fixed stars, innumerable millions of miles from us,—yet, as our eye cannot distinguish these different distances, we acquire the habit of referring all objects beyond a moderate height to one and the same surface, namely, an imaginary spherical surface, denominated the celestial vault. Thus, the various objects represented in the diagram on next page, though differing very much in shape and diameter, would all be *projected* upon the sky alike, and compose a part, indeed, of the imaginary vault itself. The place which each object occupies is determined by lines drawn from the eye of the spectator through the extremities of the body, to meet the imaginary concave sphere. Thus, to a spectator at O, Fig 16, the several lines A B, C D, and E F, would all be projected into arches on the face of the sky, and be seen as parts of the sky itself, as represented by the lines A´ B´, C´ D´, and E´ F´. And were a body actually to move in the several directions indicated by these lines, they would appear to the spectator to describe portions of the celestial vault. Thus, even when moving through the crooked line, from *a* to *b*, a body would appear to be moving along the face of the sky, and of course in a regular curve line, from *c* to *d*.

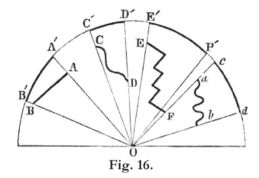

Fig. 16.

But, although all objects, beyond a certain moderate height, are projected on the imaginary surface of the sky, yet different spectators will project the same object on *different parts* of the sky. Thus, a spectator at A, Fig. 17, would see a body, C, at M, while a spectator at B would see the same body at N. This change of place in a body, as seen from different points, is called parallax, which is thus defined: *parallax is the apparent change of place which bodies undergo by being viewed from different points.*

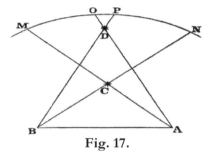

Fig. 17.

The arc M N is called the *parallactic arc*, and the angle A C B, the *parallactic angle.*

It is plain, from the figure, that near objects are much more affected by parallax than distant ones. Thus, the body C, Fig. 17, makes a much greater parallax than the more distant body D,—the former being measured by the arc M N, and the latter by the arc O P. We may easily imagine bodies to be so distant, that they would appear projected at very nearly the same point of the heavens, when viewed from places very remote from each other. Indeed, the fixed stars, as we shall see more fully hereafter, are so distant, that spectators, a hundred millions of miles apart, see each star in one and the same place in the heavens.

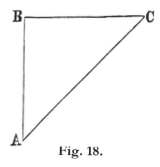

Fig. 18.

It is by means of parallax, that astronomers find the distances and magnitudes of the heavenly bodies. In order fully to understand this subject, one requires to know something of trigonometry, which science enables us to find certain unknown parts of a triangle from certain other parts which are known. Although you may not be acquainted with the principles of trigonometry, yet you will readily understand, from your knowledge of arithmetic, that from certain things given in a problem others may be found. Every triangle has of course three sides and three angles; and, if we know two of the angles and one of the sides, we can find all the other parts, namely, the remaining angle and the two unknown sides. Thus, in the triangle A B C, Fig. 18, if we know the length of the side A B, and how many degrees each of the angles A B C and B C A contains, we can find the length of the side B C, or of the side A C, and the remaining angle at A. Now, let us apply these principles to the measurements of some of the heavenly bodies.

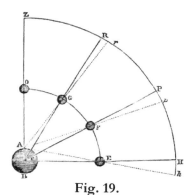

Fig. 19.

In Fig. 19, let A represent the earth, C H the horizon, and H Z a quadrant of a great circle of the heavens, extending from the horizon to the zenith; and let E, F, G, O, be successive positions of the moon, at different elevations, from the horizon to the meridian. Now, a spectator on the surface of the earth, at A, would refer the moon, when at E, to *h*, on the face of the sky, whereas, if seen from the centre of the earth, it would appear at H. So, when the moon was at F, a spectator at A would see it at *p*, while, if seen from the centre, it would have appeared at P. The parallactic arcs, H *h*, P *p*, R *r*, grow continually smaller and smaller, as a body is situated higher above the horizon; and when the body is in the zenith, then the parallax vanishes altogether, for at O the moon would be seen at Z, whether viewed from A or C.

Since, then, a heavenly body is liable to be referred to different points on the celestial vault, when seen from different parts of the earth, and thus some confusion be occasioned in the determination of points on the celestial sphere, astronomers have agreed to consider the true place of a celestial object to be that where it would appear, if seen from the centre of the earth; and the doctrine of parallax teaches how to reduce observations made at any place on the surface of the earth, to such as they would be, if made from the centre.

When the moon, or any heavenly body, is seen in the horizon, as at E, the change of place is called the horizontal parallax. Thus, the angle A E C, measures the horizontal parallax of the moon. Were a spectator to view the earth from the centre of the moon, he would see the semidiameter of the earth under this same angle; hence, *the horizontal parallax of any body is the angle subtended by the semidiameter of the earth, as seen from the body*. Please to remember this fact.

It is evident from the figure, that the effect of parallax upon the place of a celestial body is to *depress* it. Thus, in consequence of parallax, E is depressed by the arc H *h*; F, by the arc P *p*; G, by the arc R *r*; while O sustains no change. Hence, in all calculations respecting the altitude of the sun, moon, or planets, the amount of parallax is to be added: the stars, as we shall see hereafter, have no sensible parallax.

It is now very easy to see how, when the parallax of a body is known, we may find its distance from the centre of the earth. Thus, in the triangle A C E, Fig. 19, the side A C is known, being the semidiameter of the earth; the angle C A E, being a right angle, is also known; and the parallactic angle, A E C, is found from observation; and it is a well-known principle of trigonometry, that when we have any two angles of a triangle, we may find the remaining angle by subtracting the sum of these two from one hundred and eighty degrees. Consequently, in the triangle A E C, we know all the angles and one side, namely, the side A C; hence, we have the means of finding the side C E, which is the distance from the centre of the earth to the centre of the moon.

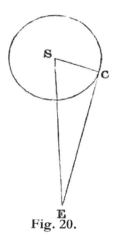

Fig. 20.

When the distance of a heavenly body is known, and we can measure, with instruments, its angular breadth, we can easily determine its *magnitude*. Thus, if we have the distance of the moon, E S, Fig. 20, and half the breadth of its disk S C, (which is measured by the angle S E C,) we can find the length of the line, S C, in miles. Twice this line is the diameter of the body; and when we know the diameter of a sphere, we can, by well-known rules, find the contents of the surface, and its solidity.

You will perhaps be curious to know, *how the moon's horizontal parallax is found*; for it must have been previously ascertained, before we could apply this method to finding the distance of the moon from the earth. Suppose that two astronomers take their stations on the same meridian, but one south of the equator, as at the Cape of Good Hope, and another north of the equator, as at Berlin, in Prussia, which two places lie nearly on the same meridian. The observers would severally refer the moon to different points on the face of the sky,—the southern observer carrying it further north, and the northern observer further south, than its true place, as seen from the centre of the earth. This will be plain from the diagram, Fig. 21. If A and B represent the positions of the spectators, M the moon, and C D an arc of the sky, then it is evident, that C D would be the parallactic arc.

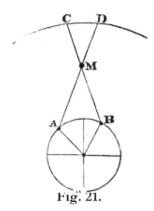

Fig. 21.

These observations furnish materials for calculating, by the aid of trigonometry, the moon's horizontal parallax, and we have before seen how, when we know the parallax of a heavenly body, we can find both its distance from the earth and its magnitude.

Beside the change of place which these heavenly bodies undergo, in consequence of parallax, there is another, of an opposite kind, arising from the effect of the atmosphere, called *refraction*. Refraction elevates the apparent place of a body, while parallax depresses it. It affects alike the most distant as well as nearer bodies.

In order to understand the nature of refraction, we must consider, that an object always appears in the direction in which the *last* ray of light comes to the eye. If the light which comes from a star were bent into fifty directions before it reached the eye, the star would nevertheless appear in the line described by the ray nearest the eye. The operation of this principle is seen when an oar, or any stick, is thrust into water. As the rays of light by which the oar is seen have their direction changed as they pass out of water into air, the apparent direction in which the body is seen is changed in the same degree, giving it a bent appearance,—the part below the water having apparently a different direction from the part above. Thus, in Fig. 22, page 96, if S *a x* be the oar, S *a b* will be the bent appearance, as affected by refraction. The transparent substance through which any ray of light passes is called a *medium*. It is a general fact in optics, that, when light passes out of a rarer into a denser medium, as out of air into water, or out of space into air, it is turned *towards* a perpendicular to the surface of the medium; and when it passes out of a denser into a rarer medium, as out of water into air, it is turned *from* the perpendicular. In the above case, the light, passing out of space into air, is turned towards the radius of the earth, this being perpendicular to the surface of the atmosphere; and it is turned more and more towards that radius the nearer it approaches to the earth, because the density of the air rapidly increases near the earth.

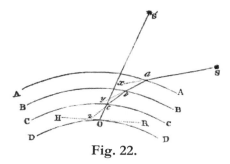

Fig. 22.

Let us now conceive of the atmosphere as made up of a great number of parallel strata, as A A, B B, C C, and D D, increasing rapidly in density (as is known to be the fact) in approaching near to the surface of the earth. Let S be a star, from which a ray of light, S a, enters the atmosphere at a, where, being much turned towards the radius of the convex surface, it would change its direction into the line a b, and again into b c, and c O, reaching the eye at O. Now, since an object always appears in the direction in which the light finally strikes the eye, the star would be seen in the direction O c, and, consequently, the star would apparently change its place, by refraction, from S to S´, being elevated out of its true position. Moreover, since, on account of the continual increase of density in descending through the atmosphere, the light would be continually turned out of its course more and more, it would therefore move, not in the polygon represented in the figure, but in a corresponding curve line, whose curvature is rapidly increased near the surface of the earth.

When a body is in the zenith, since a ray of light from it enters the atmosphere at right angles to the refracting medium, it suffers no refraction. Consequently, the position of the heavenly bodies, when in the zenith, is not changed by refraction, while, near the horizon, where a ray of light strikes the medium very obliquely, and traverses the atmosphere through its densest part, the refraction is greatest. The whole amount of refraction, when a body is in the horizon, is thirty-four minutes; while, at only an elevation of one degree, the refraction is but twenty-four minutes; and at forty-five degrees, it is scarcely a single minute. Hence it is always important to make our observations on the heavenly bodies when they are at as great an elevation as possible above the horizon, being then less affected by refraction than at lower altitudes.

Since the whole amount of refraction near the horizon exceeds thirty-three minutes, and the diameters of the sun and moon are severally less than this, these luminaries are in view both before they have actually risen and after they have set.

The rapid increase of refraction near the horizon is strikingly evinced by the *oval* figure which the sun assumes when near the horizon, and which is seen to the greatest advantage when light clouds enable us to view the solar disk. Were all parts of the sun equally raised by refraction, there would be no change of figure; but, since the lower side is more refracted than the upper, the effect is to shorten the vertical diameter, and thus to give the disk an oval form. This effect is particularly remarkable when the sun, at his rising or setting, is observed from the top of a mountain, or at an elevation near the seashore; for in such situations, the rays of light make a greater angle than ordinary with a perpendicular to the refracting medium, and the amount of refraction is proportionally greater. In some cases of this kind, the shortening of the vertical diameter of the sun has been observed to amount to six minutes, or about one fifth of the whole.

The apparent enlargement of the sun and moon, when near the horizon, arises from an optical illusion. These bodies, in fact, are not seen under so great an angle when in the horizon as when on the meridian, for they are nearer to us in the latter case than in the former. The distance of the sun, indeed, is so great, that it makes very little difference in his apparent diameter whether he is viewed in the horizon or on the meridian; but with the moon, the case is otherwise; its angular diameter, when measured with instruments, is perceptibly larger when at its culmination, or highest elevation above the horizon. Why, then, do the sun and moon appear so much larger when near the horizon? It is owing to a habit of the mind, by which we judge of the magnitudes of distant objects, not merely by the angle they subtend at the eye, but also by our impressions respecting their distance, allowing, under a given angle, a greater magnitude as we imagine the distance of a body to be greater. Now, on account of the numerous objects usually in sight between us and the sun, when he is near the horizon, he appears much further removed from us than when on the meridian; and we unconsciously assign to him a proportionally greater magnitude. If we view the sun, in the two positions, through a smoked glass, no such difference of size is observed; for here no objects are seen but the sun himself.

Twilight is another phenomenon depending on the agency of the earth's atmosphere. It is that illumination of the sky which takes place just before sunrise and which continues after sunset. It is owing partly to refraction, and partly to reflection, but mostly to the latter. While the sun is within eighteen degrees of the horizon, before it rises or after it sets, some portion of its light is conveyed to us, by means of numerous reflections from the atmosphere. At the equator, where the circles of daily motion are perpendicular to the horizon, the sun descends through eighteen degrees in an hour and twelve minutes. The light of day, therefore, declines rapidly, and as rapidly advances after daybreak in the morning. At the pole, a constant twilight is enjoyed while the sun is within eighteen degrees of the horizon, occupying nearly two thirds of the half year when the direct light of the sun is withdrawn, so that the progress from continual day to constant night is exceedingly gradual. To an inhabitant of an oblique sphere, the twilight is longer in proportion as the place is nearer the elevated pole.

Were it not for the power the atmosphere has of dispersing the solar light, and scattering it in various directions, no objects would be visible to us out of direct sunshine; every shadow of a passing cloud would involve us in midnight darkness; the stars would be visible all day; and every apartment into which the sun had not direct admission would be involved in the obscurity of night. This scattering action of the atmosphere on the solar light is greatly increased by the irregularity of temperature caused by the sun, which throws the atmosphere into a constant state of undulation; and by thus bringing together masses of air of different temperatures, produces partial reflections and refractions at their common boundaries, by which means much light is turned aside from a direct course, and diverted to the purposes of general illumination.[6] In the upper regions of the atmosphere, as on the tops of very high mountains, where the air is too much rarefied to reflect much light, the sky assumes a black appearance, and stars become visible in the day time.

Although the atmosphere is usually so transparent, that it is invisible to us, yet we as truly move and live in a fluid as fishes that swim in the sea. Considered in comparison with the whole earth, the atmosphere is to be regarded as a thin layer investing the surface, like a film of water covering the surface of an orange. Its actual height, however, is over a hundred miles, though we cannot assign its precise boundaries. Being perfectly elastic, the lower portions, bearing as they do, the weight of all the mass above them, are greatly compressed, while the upper portions having little to oppose the natural tendency of air to expand, diffuse themselves widely. The consequence is, that the atmosphere undergoes a rapid diminution of density, as we ascend from the earth, and soon becomes exceedingly rare. At so moderate a height as seven miles, it is four times rarer than at the surface, and continues to grow rare in the same proportion, namely, being four times less for every seven miles of ascent. It is only, therefore, within a few miles of the earth, that the atmosphere is sufficiently dense to sustain clouds and vapors, which seldom rise so high as eight miles, and are usually much nearer to the earth than this. So rare does the air become on the top of Mount Chimborazo, in South America, that it is incompetent to support most of the birds that fly near the level of the sea. The condor, a bird which has remarkably long wings, and a light body, is the only bird seen towering above this lofty summit. The transparency of the atmosphere,—a quality so essential to fine views of the starry heavens,—is much increased by containing a large proportion of water, provided it is perfectly dissolved, or in a state of invisible vapor. A country at once hot and humid, like some portions of the torrid zone, presents a much brighter and more beautiful view of the moon and stars, than is seen in cold climates. Before a copious rain, especially in hot weather, when the atmosphere is unusually humid, we sometimes observe the sky to be remarkably resplendent, even in our own latitude. Accordingly, this unusual clearness of the sky, when the stars shine with unwonted brilliancy, is regarded as a sign of approaching rain; and when, after the rain is apparently over, the air is remarkably transparent, and distant objects on the earth are seen with uncommon distinctness, while the sky exhibits an unusually deep azure, we may conclude that the serenity is only temporary, and that the rain will probably soon return.

LETTER X.

THE SUN.

"Great source of day! best image here below
Of thy Creator, ever pouring wide,
From world to world, the vital ocean round,
On Nature write, with every beam, His praise."—*Thomson*.

THE subjects which have occupied the preceding Letters are by no means the most interesting parts of our science. They constitute, indeed, little more than an introduction to our main subject, but comprise such matters as are very necessary to be clearly understood, before one is prepared to enter with profit and delight upon the more sublime and interesting field which now opens before us.

We will begin our survey of the heavenly bodies with the SUN, which first claims our homage, as the natural monarch of the skies. The moon will next occupy our attention; then the other bodies which compose the solar system, namely, the planets and comets; and, finally, we shall leave behind this little province in the great empire of Nature, and wing a bolder flight to the fixed stars.

The *distance* of the sun from the earth is about ninety-five millions of miles. It may perhaps seem incredible to you, that astronomers should be able to determine this fact with any degree of certainty. Some, indeed, not so well informed as yourself, have looked upon the marvellous things that are told respecting the distances, magnitudes, and velocities, of the heavenly bodies, as attempts of astronomers to impose on the credulity of the world at large; but the certainty and exactness with which the predictions of astronomers are fulfilled, as an eclipse, for example, ought to inspire full confidence in their statements. I can assure you, my dear friend, that the evidence on which these statements are founded is perfectly satisfactory to those whose attainments in the sciences qualify them to understand them; and, so far are astronomers from wishing to impose on the unlearned, by announcing such wonderful discoveries as they have made among the heavenly bodies, no class of men have ever shown a stricter regard and zeal than they for the exact truth, wherever it is attainable.

Ninety-five millions of miles is indeed a vast distance. No human mind is adequate to comprehend it fully; but the nearest approaches we can make towards it are gained by successive efforts of the mind to conceive of great distances, beginning with such as are clearly within our grasp. Let us, then, first take so small a distance as that of the breadth of the Atlantic ocean, and follow, in mind, a ship, as she leaves the port of New York, and, after twenty days' steady sail, reaches Liverpool. Having formed the best idea we are able of this distance, we may then reflect, that it would take a ship, moving constantly at the rate of ten miles per hour, more than a thousand years to reach the sun.

The diameter of the sun is towards a million of miles; or, more exactly, it is eight hundred and eighty-five thousand miles. One hundred and twelve bodies as large as the earth, lying side by side, would be required to reach across the solar disk; and our ship, sailing at the same rate as before, would be ten years in passing over the same space. Immense as is the sun, we can readily understand why it appears no larger than it does, when we reflect, that its distance is still more vast. Even large objects on the earth, when seen on a distant eminence, or over a wide expanse of water, dwindle almost to a point. Could we approach nearer and nearer to the sun, it would constantly expand its volume, until finally it would fill the whole vault of heaven. We could, however, approach but little nearer to the sun without being consumed by the intensity of his heat. Whenever we come nearer to any fire, the heat rapidly increases, being four times as great at half the distance, and one hundred times as great at one tenth the distance. This fact is expressed by saying, that the heat increases as the square of the distance decreases. Our globe is situated at such a distance from the sun, as exactly suits the animal and vegetable kingdoms. Were it either much nearer or much more remote, they could not exist, constituted as they are. The intensity of the solar light also follows the same law. Consequently, were we nearer to the sun than we are, its blaze would be insufferable; or, were we much further off, the light would be too dim to serve all the purposes of vision.

The sun is one million four hundred thousand times as large as the earth; but its matter is not more than about one fourth as dense as that of the earth, being only a little heavier than water, while the average density of the earth is more than five times that of water. Still, on account of the immense magnitude of the sun, its entire quantity of matter is three hundred and fifty thousand times as great as that of the earth. Now, the force of gravity in a body is greater, in proportion as its quantity of matter is greater. Consequently, we might suppose, that the weight of a body (weight being nothing else than the measure of the force of gravity) would be increased in the same proportion; or, that a body, which weighs only one pound at the surface of the earth, would weigh three hundred and fifty thousand pounds at the sun. But we must consider, that the attraction exerted by any body is the same as though all the matter were concentrated in the centre. Thus, the attraction exerted by the earth and by the sun is the same as though the entire matter of each body were in its centre. Hence, on account of the vast dimensions of the sun, its surface is one hundred and twelve times further from its centre than the surface of the earth is from its centre; and, since the force of gravity diminishes as the square of the distance increases, that of the sun, exerted on bodies at its surface, is (so far as this cause operates) the square of one hundred and twelve, or twelve thousand five hundred and forty-four times less than that of the earth. If, therefore, we increase the weight of a body three hundred and fifty-four thousand times, in consequence of the greater amount of matter in the sun, and diminish it twelve thousand five hundred and forty-four times, in consequence of the force acting at a greater distance from the body, we shall find that the body would weigh about twenty-eight times more on the sun than on the earth. Hence, a man weighing three hundred pounds would, if conveyed to the surface of the sun, weigh eight thousand four hundred pounds, or nearly three tons and three quarters. A limb of our bodies, weighing forty pounds, would require to lift it a force of one thousand one hundred and twenty pounds, which would be beyond the ordinary power of the muscles. At the surface of the earth, a body falls from rest by the force of gravity, in one second, sixteen and one twelfth feet; but at the surface of the sun, a body would, in the same time, fall through four hundred and forty-eight and seven tenths feet.

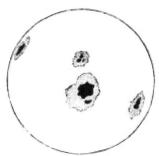

Fig. 23.

The sun turns on his own axis once in a little more than twenty-five days. This fact is known by observing certain dark places seen by the telescope on the sun's disk, called *solar spots*. These are very variable in size and number. Sometimes, the sun's disk, when viewed with a telescope, is quite free from spots, while at other times we may see a dozen or more distinct clusters, each containing a great number of spots, some large and some very minute. Occasionally, a single spot is so large as to be visible to the naked eye, especially when the sun is near the horizon, and the glare of his light is taken off. One measured by Dr. Herschel was no less than fifty thousand miles in diameter. A solar spot usually consists of two parts, the *nucleus* and the *umbra*. The nucleus is black, of a very irregular shape, and is subject to great and sudden changes, both in form and size. Spots have sometimes seemed to burst asunder, and to project fragments in different directions. The umbra is a wide margin, of lighter shade, and is often of greater extent than the nucleus. The spots are usually confined to a zone extending across the central regions of the sun, not exceeding sixty degrees in breadth. Fig. 23 exhibits the appearance of the solar spots. As these spots have all a common motion from day to day, across the sun's disk; as they go off at one limb, and, after a certain interval, sometimes come on again on the opposite limb, it is inferred that this apparent motion is imparted to them by an actual revolution of the sun on his own axis. We know that the spots must be in contact, or very nearly so, at least, with the body of the sun, and cannot be bodies revolving around it, which are projected on the solar disk when they are between us and the sun; for, in that case, they would not be so long in view as out of view, as will be evident from inspecting the following diagram. Let S, Fig. 24, page 106, represent the sun, and *b* a body revolving round it in the orbit *a b c*; and let E represent the earth, where, of course, the spectator is situated. The body would be seen projected on the sun only while passing from *b* to *c*, while, throughout the remainder of its orbit, it would be out of view, whereas no such inequality exists in respect to the two periods.

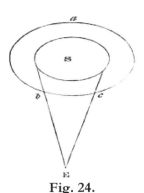

Fig. 24.

If you ask, what is the *cause* of the solar spots, I can only tell you what different astronomers have supposed respecting them. They attracted the notice of Galileo soon after the invention of the telescope, and he formed an hypothesis respecting their nature. Supposing the sun to consist of a solid body embosomed in a sea of liquid fire, he believed that the spots are composed of black cinders, formed in the interior of the sun by volcanic action, which rise and float on the surface of the fiery sea. The chief objections to this hypothesis are, first, the *vast extent* of some of the spots, covering, as they do, two thousand millions of square miles, or more,—a space which it is incredible should be filled by lava in so short a time as that in which the spots are sometimes formed; and, secondly, the *sudden disappearance* which the spots sometimes undergo, a fact which can hardly be accounted for by supposing, as Galileo did, that such a vast accumulation of matter all at once sunk beneath the fiery flood. Moreover, we have many reasons for believing that the spots are *depressions* below the general surface.

La Lande, an eminent French astronomer of the last century, held that the sun is a solid, opaque body, having its exterior diversified with high mountains and deep valleys, and covered all over with a burning sea of liquid matter. The spots he supposed to be produced by the flux and reflux of this fiery sea, retreating occasionally from the mountains, and exposing to view a portion of the dark body of the sun. But it is inconsistent with the nature of fluids, that a liquid, like the sea supposed, should depart so far from its equilibrium and remain so long fixed, as to lay bare the immense spaces occupied by some of the solar spots.

Dr. Herschel's views respecting the nature and constitution of the sun, embracing an explanation of the solar spots, have, of late years, been very generally received by the astronomical world. This great astronomer, after attentively viewing the surface of the sun, for a long time, with his large telescopes, came to the following conclusions respecting the nature of this luminary. He supposes the sun to be a planetary body like our earth, diversified with mountains and valleys, to which, on account of the magnitude of the sun, he assigns a prodigious extent, some of the mountains being six hundred miles high, and the valleys proportionally deep. He employs in his explanation no volcanic fires, but supposes two separate regions of dense clouds floating in the solar atmosphere, at different distances from the sun. The exterior stratum of clouds he considers as the depository of the sun's light and heat, while the inferior stratum serves as an awning or screen to the body of the sun itself, which thus becomes fitted to be the residence of animals. The proofs offered in support of this hypothesis are chiefly the following: first, that the appearances, as presented to the telescope, are such as accord better with the idea that the fluctuations arise from the motions of clouds, than that they are owing to the agitations of a liquid, which could not depart far enough from its general level to enable us to see its waves at so great a distance, where a line forty miles in length would subtend an angle at the eye of only the tenth part of a second; secondly, that, since clouds are easily dispersed to any extent, the great dimensions which the solar spots occasionally exhibit are more consistent with this than with any other hypothesis; and, finally, that a lower stratum of clouds, similar to those of our atmosphere, was frequently seen by the Doctor, far below the luminous clouds which are the fountains of light and heat.

Such are the views of one who had, it must be acknowledged, great powers of observation, and means of observation in higher perfection than have ever been enjoyed by any other individual; but, with all deference to such authority, I am compelled to think that the hypothesis is encumbered with very serious objections. Clouds analogous to those of our atmosphere (and the Doctor expressly asserts that his lower stratum of clouds are analogous to ours, and reasons respecting the upper stratum according to the same analogy) cannot exist in hot air; they are tenants only of cold regions. How can they be supposed to exist in the immediate vicinity of a fire so intense, that they are even dissipated by it at the distance of ninety-five millions of miles? Much less can they be supposed to be the depositories of such devouring fire, when any thing in the form of clouds, floating in our atmosphere, is at once scattered and dissolved by the accession of only a few degrees of heat. Nothing, moreover, can be imagined more unfavorable for radiating heat to such a distance, than the light, inconstant matter of which clouds are composed, floating loosely in the solar atmosphere. There is a logical difficulty in the case: it is ascribing to things properties which they are not known to possess; nay, more, which they are known not to possess. From variations of light and shade in objects seen at moderate distances on the earth, we are often deceived in regard to their appearances; and we must distrust the power of an astronomer to decide upon the nature of matter seen at the distance of ninety-five millions of miles.

I think, therefore, we must confess our ignorance of the nature and constitution of the sun; nor can we, as astronomers, obtain much more satisfactory knowledge respecting it than the common apprehension, namely, that it is an immense globe of fire. We have not yet learned what causes are in operation to maintain its undecaying fires; but our confidence in the Divine wisdom and goodness leads us to believe, that those causes are such as will preserve those fires from extinction, and at a nearly uniform degree of intensity. Any material change in this respect would jeopardize the safety of the animal and vegetable kingdoms, which could not exist without the enlivening influence of the solar heat, nor, indeed, were that heat any more or less intense than it is at present.

If we inquire whether the surface of the sun is in a state of actual combustion, like burning fuel, or merely in a state of intense ignition, like a stone heated to redness in a furnace, we shall find it most reasonable to conclude that it is in a state of ignition. If the body of the sun were composed of combustible matter and were actually on fire, the material of the sun would be continually wasting away, while the products of combustion would fill all the vast surrounding regions, and obscure the solar light. But solid bodies may attain a very intense state of ignition, and glow with the most fervent heat, while none of their material is consumed, and no clouds or fumes rise to obscure their brightness, or to impede their further emission of heat. An ignited surface, moreover, is far better adapted than flame to the radiation of heat. Flame, when made to act in contact with the surfaces of solid bodies, heats them rapidly and powerfully; but it sends forth, or *radiates*, very little heat, compared with solid matter in a high state of ignition. These various considerations render it highly probable to my mind, that the body of the sun is not in a state of actual combustion, but merely in a state of high ignition.

The solar beam consists of a mixture of several different sorts of rays. First, there are the *calorific* rays, which afford heat, and are entirely distinct from those which afford light, and may be separated from them. Secondly, there are the *colorific* rays, which give light, consisting of rays of seven distinct colors, namely, violet, indigo, blue, green, yellow, orange, red. These, when separated, as they may be by a glass prism, compose the *prismatic spectrum*. They appear also in the rainbow. When united again, in due proportions, they constitute white light, as seen in the light of the sun. Thirdly, there are found in the solar beam a class of rays which afford neither heat nor light, but which produce chemical changes in certain bodies exposed to their influence, and hence are called *chemical* rays. Fourthly, there is still another class, called *magnetizing* rays, because they are capable of imparting magnetic properties to steel. These different sorts of rays are sent forth from the sun, to the remotest regions of the planetary worlds, invigorating all things by their life-giving influence, and dispelling the darkness that naturally fills all space.

But it was not alone to give heat and light, that the sun was placed in the firmament. By his power of attraction, also, he serves as the great regulator of the planetary motions, bending them continually from the straight line in which they tend to move, and compelling them to circulate around him, each at nearly a uniform distance, and all in perfect harmony. I will hereafter explain to you the manner in which the gravity of the sun thus acts, to control the planetary motions. For the present, let us content ourselves with reflecting upon the wonderful force which the sun must put forth, in order to bend out of their courses, into circular orbits, such a number of planets, some of which are more than a thousand times as large as the earth. Were a ship of war under full sail, and it should be required to turn her aside from her course by a rope attached to her bow, we can easily imagine that it would take a great force to do it, especially were it required that the force should remain stationary and the ship be so constantly diverted from her course, as to be made to go round the force as round a centre. Somewhat similar to this is the action which the sun exerts on each of the planets by some invisible influence, called gravitation. The bodies which he thus turns out of their course, and bends into a circular orbit around himself, are, however, many millions of times as ponderous as the ship, and are moving many thousand times as swiftly.

LETTER XI.

ANNUAL REVOLUTION.—SEASONS

"These, as they change, Almighty Father, these
Are but the varied God. The rolling year
Is full of Thee."—*Thomson.*

WE have seen that the apparent revolution of the heavenly bodies, from east to west, every twenty-four hours, is owing to a real revolution of the earth on its own axis, in the opposite direction. This motion is very easily understood, resembling, as it does, the spinning of a top. We must, however, conceive of the top as turning without any visible support, and not as resting in the usual manner on a plane. The annual motion of the earth around the sun, which gives rise to an apparent motion of the sun around the earth once a year, and occasions the change of seasons, is somewhat more difficult to understand; and it may cost you some reflection, before you will settle all the points respecting the changes of the seasons clearly in your mind. We sometimes see these two motions exemplified in a top. When, as the string is pulled, the top is thrown forwards on the floor, we may see it move forward (sometimes in a circle) at the same time that it spins on its axis. Let a candle be placed on a table, to represent the sun, and let these two motions be imagined to be given to a top around it, and we shall have a case somewhat resembling the actual motions of the earth around the sun.

When bodies are at such a distance from each other as the earth and the sun, a spectator on either would project the other body upon the concave sphere of the heavens, always seeing it on the opposite side of a great circle one hundred and eighty degrees from himself.

Recollect that the path in which the earth moves round the sun is called the ecliptic. We are not to conceive of this, or of any other celestial circle, as having any real, palpable existence, any more than the path of a bird through the sky. You will perhaps think it quite superfluous for me to remind you of this; but, from the habit of seeing the orbits of the heavenly bodies represented in diagrams and orreries, by palpable lines and circles, we are apt inadvertently to acquire the notion, that the orbits of the planets, and other representations of the artificial sphere, have a real, palpable existence in Nature; whereas, they denote the places where mere geometrical or imaginary lines run. You might have expected to see an orrery, exhibiting a view of the sun and planets, with their various motions, particularly described in my Letter on astronomical instruments and apparatus. I must acknowledge, that I entertain a very low opinion of the utility of even the best orreries, and I cannot recommend them as auxiliaries in the study of astronomy. The numerous appendages usually connected with them, some to support them in a proper position, and some to communicate to them the requisite motions, enter into the ideas which the learner forms respecting the machinery of the heavens; and it costs much labor afterwards to divest the mind of such erroneous impressions. Astronomy can be exhibited much more clearly and beautifully to the mental eye than to the visual organ. It is much easier to conceive of the sun existing in boundless space, and of the earth as moving around him at a great distance, the mind having nothing in view but simply these two bodies, than it is, in an orrery, to contemplate the motion of a ball representing the earth, carried by a complicated apparatus of wheels around another ball, supported by a cylinder or wire, to represent the sun. I would advise you, whenever it is practicable, to think how things are in Nature, rather than how they are represented by art. The machinery of the heavens is much simpler than that of an orrery.

In endeavoring to obtain a clear idea of the revolution of the earth around the sun, imagine to yourself a plane (a geometrical plane, having merely length and breadth, but no thickness) passing through the centres of the sun and the earth, and extended far beyond the earth till it reaches the firmament of stars. Although, indeed, no such dome actually exists as that under which we figure to ourselves the vault of the sky, yet, as the fixed stars appear to be set in such a dome, we may imagine that the circles of the sphere, when indefinitely enlarged, finally reach such an imaginary vault. All that is essential is, that we should imagine this to exist far beyond the bounds of the solar system, the various bodies that compose the latter being situated close around the sun, at the centre.

Along the line where this great circle meets the starry vault, are situated a series of constellations,—Aries, Taurus, Gemini, &c.,—which occupy successively this portion of the heavens. When bodies are at such a distance from each other as the sun and the earth, I have said that a spectator on either would project the other body upon the concave sphere of the heavens, always seeing it on the opposite side of a great circle one hundred and eighty degrees from himself. The place of a body, when viewed from any point, is denoted by the position it occupies among the stars. Thus, in the diagram, Fig. 25, page 114, when the earth arrives at E, it is said to be in Aries, because, if viewed from the sun, it would be projected on that part of the heavens; and, for the same reason, to a spectator at E, the sun would be in Libra. When the earth shifts its position from Aries to Taurus, as we are unconscious of our own motion, the sun it is that appears to move from

Libra to Scorpio, in the opposite part of the heavens. Hence, as we go forward, in the order of the signs, on one side of the ecliptic, the sun seems to be moving forward at the same rate on the opposite side of the same great circle; and therefore, although we are unconscious of our own motion, we can read it, from day to day, in the motions of the sun. If we could see the stars at the same time with the sun, we could actually observe, from day to day, the sun's progress through them, as we observe the progress of the moon at night; only the sun's rate of motion would be nearly fourteen times slower than that of the moon. Although we do not see the stars when the sun is present, we can observe that it makes daily progress eastward, as is apparent from the constellations of the zodiac occupying, successively, the western sky immediately after sunset, proving that either all the stars have a common motion westward, independent of their diurnal motion, or that the sun has a motion past them from west to east. We shall see, hereafter, abundant evidence to prove, that this change in the relative position of the sun and stars, is owing to a change in the apparent place of the sun, and not to any change in the stars.

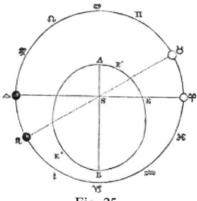

Fig. 25.

To form a clear idea of the two motions of the earth, imagine yourself standing on a circular platform which turns slowly round its centre. While you are carried slowly round the entire of the circuit of the heavens, along with the platform, you may turn round upon your heel the same way three hundred and sixty-five times. The former is analogous to our annual motion with the earth around the sun; the latter, to our diurnal revolution in common with the earth around its own axis.

Although the apparent revolution of the sun is in a direction opposite to the real motion of the earth, as regards absolute space, yet both are nevertheless from west to east, since these terms do not refer to any directions in absolute space, but to the order in which certain constellations (the constellations of the Zodiac) succeed one another. The earth itself, on opposite sides of its orbit, does in fact move towards directly opposite points of space; but it is all the while pursuing its course in the order of the signs. In the same manner, although the earth turns on its axis from west to east, yet any place on the surface of the earth is moving in a direction in space exactly opposite to its direction twelve hours before. If the sun left a visible trace on the face of the sky, the ecliptic would of course be distinctly marked on the celestial sphere, as it is on an artificial globe; and were the equator delineated in a similar manner, we should then see, at a glance, the relative position of these two circles,—the points where they intersect one another, constituting the equinoxes; the points where they are at the greatest distance asunder, that is, the solstices; and various other particulars, which, for want of such visible traces, we are now obliged to search for by indirect and circuitous methods. It will aid you, to have constantly before your mental vision an imaginary delineation of these two important circles on the face of the sky.

The equator makes an angle with the ecliptic of twenty-three degrees and twenty-eight minutes. This is called the obliquity of the ecliptic. As the sun and earth are both always in the ecliptic, and as the motion of the earth in one part of it makes the sun appear to move in the opposite part, at the same rate, the sun apparently descends, in Winter, twenty-three degrees and twenty-eight minutes to the south of the equator, and ascends, in Summer, the same number of degrees north of it. We must keep in mind, that the celestial equator and celestial ecliptic are here understood, and we may imagine them to be two great circles delineated on the face of the sky. On comparing observations made at different periods, for more than two thousand years, it is found, that the obliquity of the ecliptic is not constant, but that it undergoes a slight diminution, from age to age, amounting to fifty-two seconds in a century, or about half a second annually. We might apprehend that, by successive approaches to each other, the equator and ecliptic would finally coincide; but astronomers have discovered, by a most profound investigation, based on the principles of universal gravitation, that this irregularity is confined within certain narrow limits; and that the obliquity, after diminishing for some thousands of years, will then increase for a similar period, and will thus vibrate forever about a mean value.

As the earth traverses every part of her orbit in the course of a year, she will be once at each solstice, and once at each equinox. The best way of obtaining a correct idea of her two motions is, to conceive of her as standing still for a single day, at some point in her orbit, until she has turned once on her axis, then moving about a degree, and halting again, until another diurnal revolution is completed. Let us suppose the earth at the Autumnal equinox, the sun, of course, being at the Vernal equinox,—

for we must always think of these two bodies as diametrically opposite to each other. Suppose the earth to stand still in its orbit for twenty-four hours. The revolution of the earth on its axis, from west to east, will make the sun appear to describe a great circle of the heavens from east to west, coinciding with the equator. At the end of this period, suppose the sun to move northward one degree, and to remain there for twenty-four hours; in which time, the revolution of the earth, will make the sun appear to describe another circle, from east to west, parallel to the equator, but one degree north of it. Thus, we may conceive of the sun as moving one degree north, every day, for about three months, when it will reach the point of the ecliptic furthest from the equator, which point is called the *tropic*, from a Greek word, signifying *to turn*; because, after the sun has passed this point, his motion in his orbit carries him continually towards the equator, and therefore he seems to turn about. The same point is also called the *solstice*, from a Latin word, signifying to *stand still*; since, when the sun has reached its greatest northern or southern limit, while its declination is at the point where it ceases to increase, but begins to decrease, there the sun seems for a short time stationary, with regard to the equator, appearing for several days to describe the same parallel of latitude.

When the sun is at the northern tropic, which happens about the twenty-first of June, his elevation above the southern horizon at noon is the greatest in the year; and when he is at the southern tropic, about the twenty-first of December, his elevation at noon is the least in the year. The difference between these two meridian altitudes will give the whole distance from one tropic to the other, and consequently, twice the distance from each tropic to the equator. By this means, we find how far the tropic is from the equator, and that gives us the angle which the equator and ecliptic make with each other; for the greatest distance between any two great circles on the sphere is always equal to the angle which they make with each other. Thus, the ancient astronomers were able to determine the obliquity of the ecliptic with a great degree of accuracy. It was easy to find the situation of the zenith, because the direction of a plumb-line shows us where that is; and it was easy to find the distances from the zenith where the sun was at the greatest and least distances; respectively. The difference of these two arcs is the angular distance from one tropic to the other; and half this arc is the distance of either tropic from the equator, and of course, equal to the obliquity of the ecliptic. All this will be very easily understood from the annexed diagram, Fig. 26. Let Z be the zenith of a spectator situated at C; Z *n* the least, and Z *s* the greatest distance of the sun from the zenith. From Z *s* subtract Z *n*, and then *s n*, the difference, divided by two, will give the obliquity of the ecliptic.

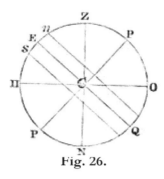

Fig. 26.

The motion of the earth in its orbit is nearly seventy times as great as its greatest motion around its axis. In its revolution around the sun, the earth moves no less than one million six hundred and forty thousand miles per day, sixty-eight thousand miles per hour, eleven hundred miles per minute, and nearly nineteen miles every second; a velocity nearly sixty times as great as the greatest velocity of a cannon ball. Places on the earth turn with very different degrees of velocity in different latitudes. Those near the equator are carried round on the circumference of a large circle; those towards the poles, on the circumference of a small circle; while one standing on the pole itself would not turn at all. Those who live on the equator are carried about one thousand miles an hour. In our latitude, (forty-one degrees and eighteen minutes,) the diurnal velocity is about seven hundred and fifty miles per hour. It would seem, at first view, quite incredible, that we should be whirled round at so rapid a rate, and yet be entirely insensible of any motion; and much more, that we could be going so swiftly through space, in our circuit around the sun, while all things, when unaffected by local causes, appear to be in such a state of quiescence. Yet we have the most unquestionable evidence of the fact; nor is it difficult to account for it, in consistency with the general state of repose among bodies on the earth, when we reflect that their relative motions, with respect to each other, are not in the least disturbed by any motions which they may have in common. When we are on board a steam-boat, we move about in the same manner when the boat is in rapid motion, as when it is lying still; and such would be the case, if it moved steadily a hundred times faster than it does. Were the earth, however, suddenly to stop its diurnal revolution, all movable bodies on its surface would be thrown off in tangents to the surface with velocities proportional to that of their diurnal motion; and were the earth suddenly to halt in its orbit, we should be hurled forward into space with inconceivable rapidity.

I will next endeavor to explain to you the phenomena of the *Seasons*. These depend on two causes; first, the inclination of the earth's axis to the plane of its orbit; and, secondly, to the circumstance, that the axis always remains parallel to itself. Imagine to yourself a candle placed in the centre of a ring, to represent the sun in the centre of the earth's orbit, and an apple with a knittingneedle running through it in the direction of the stem. Run a knife around the central part of the apple, to mark the situation of the equator. The circumference of the ring represents the earth's orbit in the plane of the ecliptic. Place the apple so that the equator shall coincide with the wire; then the axis will lie directly across the plane of the ecliptic; that is, at right angles to it. Let the apple be carried quite round the ring, constantly preserving the axis parallel to itself, and the equator all the while coinciding with the wire that represents the orbit. Now, since the sun enlightens half the globe at once, so the candle, which here represents the sun, will shine on the half of the apple that is turned towards it; and the circle which divides the enlightened from the unenlightened side of the apple, called the *terminator*, will pass through both the poles. If the apple be turned slowly round on its axis, the terminator will successively pass over all places on the earth, giving the appearance of sunrise to places at which it arrives, and of sunset to places from which it departs. If, therefore, the equator had coincided with the ecliptic, as would have been the case, had the earth's axis been perpendicular to the plane of its orbit, the diurnal motion of the sun would always have been in the equator, and the days and nights would have been equal all over the globe. To the inhabitants of the equatorial parts of the earth, the sun would always have appeared to move in the prime vertical, rising directly in the east, passing through the zenith at noon, and setting in the west. In the polar regions, the sun would always have appeared to revolve in the horizon; while, at any place between the equator and the pole, the course of the sun would have been oblique to the horizon, but always oblique in the same degree. There would have been nothing of those agreeable vicissitudes of the seasons which we now enjoy; but some regions of the earth would have been crowned with perpetual spring, others would have been scorched with the unremitting fervor of a vertical sun, while extensive regions towards either pole would have been consigned to everlasting frost and sterility.

To understand, then, clearly, the causes of the change of seasons, use the same apparatus as before; but, instead of placing the axis of the earth at right angles to the plane of its orbit, turn it out of a perpendicular position a little, (twenty-three degrees and twenty-eight minutes,) then the equator will be turned just the same number of degrees out of a coincidence with the ecliptic. Let the apple be carried around the ring, always holding the axis inclined at the same angle to the plane of the ring, and always parallel to itself. You will find that there will be two points in the circuit where the plane of the equator, that you had marked around the centre of the apple, will pass through the centre of the sun; these will be the points where the celestial equator and the ecliptic cut one another, or the equinoxes. When the earth is at either of these points, the sun shines on both poles alike; and, if we conceive of the earth, while in this situation, as turning once round on its axis, the apparent diurnal motion of the sun will be the same as it would be, were the earth's axis perpendicular to the plane of the equator. For that day, the sun would revolve in the equator, and the days and nights would be equal all over the globe. If the apple were carried round in the manner supposed, then, at the distance of ninety degrees from the equinoxes, the same pole would be turned from the sun on one side, just as much as it was turned towards him on the other. In the former case, the sun's light would fall short of the pole twenty-three and one half degrees, and in the other case, it would reach beyond it the same number of degrees. I would recommend to you to obtain as clear an idea as you can of the cause of the change of seasons, by thinking over the foregoing illustration. You may then clear up any remaining difficulties, by studying the diagram, Fig. 27, on page 122.

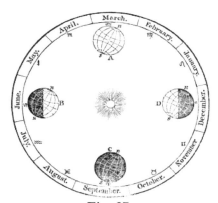

Fig. 27.

Let A B C D represent the earth's place in different parts of its orbit, having the sun in the centre. Let A, C, be the positions of the earth at the equinoxes, and B, D, its positions at the tropics,—the axis *n s* being always parallel to itself. It is difficult to represent things of this kind correctly, all on the same plane; but you will readily see, that the figure of the earth, here, answers to the apple in the former illustration; that the hemisphere towards *n* is above, and that towards *s* is below, the plane of the paper. When the earth is at A and C, the Vernal and Autumnal equinoxes, the sun, you will perceive, shines on both the poles *n* and *s*; and, if you conceive of the globe, while in this position, as turned round on its axis, as it is in the diurnal revolution, you will readily understand, that the sun would describe the celestial equator. This may not at first appear so obvious, by inspecting the figure; but if you consider the point *n* as raised above the plane of the paper, and the point *s* as depressed below it, you will readily see how the plane of the equator would pass through the centre of the sun. Again, at B, when the earth is at the southern tropic, the sun shines twenty-three and a half degrees beyond the north pole, *n*, and falls the same distance short of the south pole, *s*. The case is exactly reversed when the earth is at the northern tropic, and the sun at the southern. While the earth is at one of the tropics, at B, for example, let us conceive of it as turning on its axis, and we shall readily see, that all that part of the earth which lies within the north polar circle will enjoy continual day, while that within the south polar circle will have continual night; and that all other places will have their days longer as they are nearer to the enlightened pole, and shorter as they are nearer to the unenlightened pole. This figure likewise shows the successive positions of the earth, at different periods of the year, with respect to the signs, and what months correspond to particular signs. Thus, the earth enters Libra, and the sun Aries, on the twenty-first of March, and on the twenty-first of June, the earth is just entering Capricorn, and the sun, Cancer. You will call to mind what is meant by this phraseology,—that by saying the earth enters Libra, we mean that a spectator placed on the sun would see the earth in that part of the celestial ecliptic, which is occupied by the sign Libra; and that a spectator on the earth sees the sun at the same time projected on the opposite part of the heavens, occupied by the sign Cancer.

Had the axis of the earth been perpendicular to the plane of the ecliptic, then the sun would always have appeared to move in the equator, the days would every where have been equal to the nights, and there could have been no change of seasons. On the other hand, had the inclination of the ecliptic to the equator been much greater than it is, the vicissitudes of the seasons would have been proportionally greater, than at present. Suppose, for instance, the equator had been at right angles to the ecliptic, in which case, the poles of the earth would have been situated in the ecliptic itself; then, in different parts of the earth, the appearances would have been as follows: To a spectator on the *equator*, (where all the circles of diurnal revolution are perpendicular to the horizon,) the sun, as he left the vernal equinox, would every day perform his diurnal revolution in a smaller and smaller circle, until he reached the north pole, when he would halt for a moment, and then wheel about and return to the equator, in a reverse order. The progress of the sun through the southern signs, to the south pole, would be similar to that already described. Such would be the appearances to an inhabitant of the equatorial regions. To a spectator living in an *oblique* sphere, in our own latitude, for example, the sun, while north of the equator, would advance continually northward, making his diurnal circuit in parallels further and further distant from the equator, until he reached the circle of perpetual apparition; after which, he would climb, by a spiral course, to the north star, and then as rapidly return to the equator. By a similar progress southward, the sun would at length pass the circle of perpetual occultation, and for some time (which would be longer or shorter, according to the latitude of the place of observation) there would be continual night. To a spectator on the *pole* of the earth and under the pole of the heaven, during the long day of six months, the sun would wind its way to a point directly over head, pouring down upon the earth beneath not merely the heat of the torrid zone, but the heat of a torrid noon, accumulating without intermission.

The great vicissitudes of heat and cold, which would attend these several movements of the sun, would be wholly incompatible with the existence of either the animal or the vegetable kingdom, and all terrestrial Nature would be doomed to perpetual sterility and desolation. The happy provision which the Creator has made against such extreme vicissitudes, by confining the changes of the seasons within such narrow bounds, conspires with many other express arrangements in the economy of Nature, to secure the safety and comfort of the human race.

Perhaps you have never reflected upon all the reasons, why the several changes of position, with respect to the horizon, which the sun undergoes in the course of the year, occasion such a difference in the amount of heat received from him. Two causes contribute to increase the heat of Summer and the cold of Winter. The higher the sun ascends above the horizon, the more directly his rays fall upon the earth; and their heating power is rapidly augmented, as they approach a perpendicular direction. When the sun is nearly over head, his rays strike us with far greater force than when they meet us obliquely; and the earth absorbs a far greater number of those rays of heat which strike it perpendicularly, than of those which meet it in a slanting direction. When the sun is near the horizon, his rays merely glance along the ground, and many of them, before they reach it, are absorbed and dispersed in passing through the atmosphere. Those who have felt only the oblique solar rays, as they fall upon objects in the high latitudes, have a very inadequate idea of the power of a vertical, noonday sun, as felt in the region of the equator.

The increased length of the day in Summer is another cause of the heat of this season of the year. This cause more sensibly affects places far removed from the equator, because at such places the days are longer and the nights shorter than in the torrid zone. By the operation of this cause, the solar heat accumulates there so much, during the longest days of

Summer, that the temperature rises to a higher degree than is often known in the torrid climates.

But the temperature of a place is influenced very much by several other causes, as well as by the force and duration of the sun's heat. First, the *elevation* of a country above the level of the sea has a great influence upon its climate. Elevated districts of country, even in the torrid zone, often enjoy the most agreeable climate in the world. The cold of the upper regions of the atmosphere modifies and tempers the solar heat, so as to give a most delightful softness, while the uniformity of temperature excludes those sudden and excessive changes which are often experienced in less favored climes. In ascending certain high mountains situated within the torrid zone, the traveller passes, in a short time, through every variety of climate, from the most oppressive and sultry heat, to the soft and balmy air of Spring, which again is succeeded by the cooler breezes of Autumn, and then by the severest frosts of Winter. A corresponding difference is seen in the products of the vegetable kingdom. While Winter reigns on the summit of the mountain, its central regions may be encircled with the verdure of Spring, and its base with the flowers and fruits of Summer. Secondly, the proximity of the *ocean* also has a great effect to equalize the temperature of a place. As the ocean changes its temperature during the year much less than the land, it becomes a source of warmth to contiguous countries in Winter, and a fountain of cool breezes in Summer. Thirdly, the relative *humidity* or *dryness* of the atmosphere of a place is of great importance, in regard to its effects on the animal system. A dry air of ninety degrees is not so insupportable as a humid air of eighty degrees; and it may be asserted as a general principle, that a hot and humid atmosphere is unhealthy, although a hot air, when dry, may be very salubrious. In a warm atmosphere which is dry, the evaporation of moisture from the surface of the body is rapid, and its cooling influence affords a most striking relief to an intense heat without; but when the surrounding atmosphere is already filled with moisture, no such evaporation takes place from the surface of the skin, and no such refreshing effects are experienced from this cause. Moisture collects on the skin; a sultry, oppressive sensation is felt; and chills and fevers are usually in the train.

LETTER XII.

LAWS OF MOTION.

"What though in solemn silence, all
Move round this dark, terrestrial ball!
In reason's ear they all rejoice,
And utter forth a glorious voice;
For ever singing, as they shine,'
The hand that made us is divine.'"—*Addison.*

HOWEVER incredible it may seem, no fact is more certain, than that the earth is constantly on the wing, flying around the sun with a velocity so prodigious, that, for every breath we draw, we advance on our way forty or fifty miles. If, when passing across the waters in a steam-boat, we can wake, after a night's repose, and find ourselves conducted on our voyage a hundred miles, we exult in the triumphs of art, which could have moved so ponderous a body as a steam-ship over such a space in so short a time, and so quietly, too, as not to disturb our slumbers; but, with a motion vastly more quiet and uniform, we have, in the same interval, been carried along with the earth in its orbit more than half a million of miles. In the case of the steam-ship, however perfect the machinery may be, we still, in our waking hours at least, are made sensible of the action of the forces by which the motion is maintained,—as the roaring of the fire, the beating of the piston, and the dashing of the paddle-wheels; but in the more perfect machinery which carries the earth forward on her grander voyage, no sound is heard, nor the least intimation afforded of the stupendous forces by which this motion is achieved. To the pious observer of Nature it might seem sufficient, without any inquiry into second causes, to ascribe the motions of the spheres to the direct agency of the Supreme Being. If, however, we can succeed in finding the secret springs and cords, by which the motions of the heavenly bodies are immediately produced and controlled, it will detract nothing from our just admiration of the Great First Cause of all things. We may therefore now enter upon the inquiry into the nature or laws of the forces by which the earth is made to revolve on her axis and in her orbit; and having learned what it is, that causes and maintains the motions of the earth, you will then acquire, at the same time, a knowledge of all the celestial machinery. The subject will involve an explanation of the laws of motion, and of the principles of universal gravitation.

It was once supposed, that we could never reason respecting the laws that govern the heavenly bodies from what we observe in bodies around us, but that motion is one thing on the earth and quite another thing in the skies; and hence, that it is impossible for us, by any inquiries into the laws of terrestrial Nature, to ascertain how things take place among the heavenly bodies. Galileo and Newton, however, proceeded on the contrary supposition, that Nature is uniform in all her works; that the same Almighty arm rules over all; and that He works by the same fixed laws through all parts of His boundless realm. The certainty with which all the predictions of astronomers, made on these suppositions, are fulfilled, attests the soundness of the hypothesis. Accordingly, those laws, which all experience, endlessly multiplied and varied, proves to be the laws of terrestrial motion, are held to be the laws that govern also the motions of the most distant planets and stars, and to prevail throughout the universe of matter. Let us, then, briefly review these great laws of motion, which are three in number. The FIRST LAW is as follows: *every body perseveres in a state of rest, or of uniform motion in a straight line, unless compelled by some force to change its state.* By *force* is meant any thing which produces motion.

The foregoing law has been fully established by experiment, and is conformable to all experience. It embraces several particulars. First, a body, when at rest, remains so, unless some force puts it in motion; and hence it is inferred, when a body is found in motion, that some force must have been applied to it sufficient to have caused its motion. Thus, the fact, that the earth is in motion around the sun and around its own axis, is to be accounted for by assigning to each of these motions a force adequate, both in quantity and direction, to produce these motions, respectively.

Secondly, when a body is once in motion, it will continue to move for ever, unless something stops it. When a ball is struck on the surface of the earth, the friction of the earth and the resistance of the air soon stop its motion; when struck on smooth ice, it will go much further before it comes to a state of rest, because the ice opposes much less resistance than the ground; and, were there no impediment to its motion, it would, when once set in motion, continue to move without end. The heavenly bodies are actually in this condition: they continue to move, not because any new forces are applied to them; but, having been once set in motion, they continue in motion because there is nothing to stop them. This property in bodies to persevere in the state they are actually in,—if at rest, to remain at rest, or, if in motion, to continue in motion,—is called *inertia*. The inertia of a body (which is measured by the force required to overcome it) is proportioned to the quantity of matter it contains. A steam-boat manifests its inertia, on first starting it, by the enormous expenditure of force required to bring it to a given rate of motion; and it again manifests its inertia, when in rapid motion, by the great difficulty of stopping it. The heavenly bodies, having been once put in motion, and meeting with nothing to stop them, move on by their own inertia. A top affords a beautiful illustration of inertia, continuing, as it does, to spin after the moving force is withdrawn.

Thirdly, the motion to which a body naturally tends is *uniform*; that is, the body moves just as far the second minute as it did the first, and as far the third as the second; and passes over equal spaces in equal times. I do not assert that the motion of all moving bodies is *in fact* uniform, but that such is their *tendency*. If it is otherwise than uniform, there is some cause operating to disturb the uniformity to which it is naturally prone.

Fourthly, a body in motion will move in a *straight line*, unless diverted out of that line by some external force; and the body will resume its straight-forward motion, whenever the force that turns it aside is withdrawn. Every body that is revolving in an orbit, like the moon around the earth, or the earth around the sun, *tends* to move in a straight line which is a tangent[7] to its orbit. Thus, if A B C, Fig. 28, represents the orbit of the moon around the earth, were it not for the constant action of some force that draws her towards the earth, she would move off in a straight line. If the force that carries her towards the earth were suspended at A, she would immediately desert the circular motion, and proceed in the direction A D. In the same manner, a boy whirls a stone around his head in a sling, and then letting go one of the strings, and releasing the force that binds it to the circle, it flies off in a straight line which is a tangent to that part of the circle where it was released. This tendency which a body revolving in an orbit exhibits, to recede from the centre and to fly off in a tangent, is called the *centrifugal force*. We see it manifested when a pail of water is whirled. The water rises on the sides of the vessel, leaving a hollow in the central parts. We see an example of the effects of centrifugal action, when a horse turns swiftly round a corner, and the rider is thrown outwards; also, when a wheel passes rapidly through a small collection of water, and portions of the water are thrown off from the top of the wheel in straight lines which are tangents to the wheel.

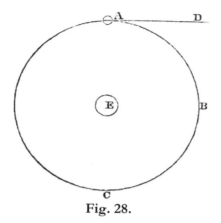
Fig. 28.

The centrifugal force is increased as the velocity is increased. Thus, the parts of a millstone most remote from the centre sometimes acquire a centrifugal force so much greater than the central parts, which move much slower, that the stone is divided, and the exterior portions are projected with great violence. In like manner, as the equatorial parts of the earth, in the diurnal revolution, revolve much faster than the parts towards the poles, so the centrifugal force is felt most at the equator, and becomes strikingly manifest by the diminished weight of bodies, since it acts in opposition to the force of gravity.

Although the foregoing law of motion, when first presented to the mind, appears to convey no new truth, but only to enunciate in a formal manner what we knew before; yet a just understanding of this law, in all its bearings, leads us to a clear comprehension of no small share of all the phenomena of motion. The second and third laws may be explained in fewer terms.

The SECOND LAW of motion is as follows: *motion is proportioned to the force impressed, and in the direction of that force.*

The meaning of this law is, that every force that is applied to a body produces its full effect, proportioned to its intensity, either in causing or in preventing motion. Let there be ever so many blows applied at once to a ball, each will produce its own effect in its own direction, and the ball will move off, not indeed in the zigzag, complex lines corresponding to the directions of the several forces, but in a single line expressing the united effect of all. If you place a ball at the corner of a table, and give it an impulse, at the same instant, with the thumb and finger of each hand, one impelling it in the direction of one side of the table, and the other in the direction of the other side, the ball will move diagonally across the table. If the blows be exactly proportioned each to the length of the side of the table on which it is directed, the ball will run exactly from corner to corner, and in the same time that it would have passed over each side by the blow given in the direction of that side. This principle is expressed by saying, that a body impelled by two forces, acting respectively in the directions of the two sides of a parallelogram, and proportioned in intensity to the lengths of the sides, will describe the diagonal of the parallelogram in the same time in which it would have described the sides by the forces acting separately.

The converse of this proposition is also true, namely, that any single motion may be considered as the *resultant* of two others,—the motion itself being represented by the diagonal, while the two *components* are represented by the sides, of a parallelogram. This reduction of a motion to the individual motions that produce it, is called the *resolution of motion*, or the *resolution of forces*. Nor can a given motion be resolved into *two* components, merely. These, again, may be resolved into others, varying indefinitely, in direction and intensity, from all which the given motion may be considered as having resulted. This composition and resolution of motion or forces is often of great use, in inquiries into the motions of the heavenly bodies. The composition often enables us to substitute a single force for a great number of others, whose individual operations would be too complicated to be followed. By this means, the investigation is greatly simplified. On the other hand, it is frequently very convenient to resolve a given motion into two or more others, some of which may be thrown out of the account, as not influencing the particular point which we are inquiring about, while others are far more easily understood and managed than the single force would have been. It is characteristic of great minds, to simplify these inquiries. They gain an insight into complicated and difficult subjects, not so much by any extraordinary faculty of seeing in the dark, as by the power of removing from the object all incidental causes of obscurity, until it shines in its own clear and simple light.

If every force, when applied to a body, produces its full and legitimate effect, how many other forces soever may act upon it, impelling it different ways, then it must follow, that the smallest force ought to move the largest body; and such is in fact the case. A snap of a finger upon a seventy-four under full sail, if applied in the direction of its motion, would actually increase its speed, although the effect might be too small to be visible. Still it is something, and may be truly expressed by a fraction. Thus, suppose a globe, weighing a million of pounds, were suspended from the ceiling by a string, and we should apply to it the snap of a finger,—it is granted that the motion would be quite insensible. Let us then divide the body into a million equal parts, each weighing one pound; then the same impulse, applied to each one separately, would produce a sensible effect, moving it, say one inch. It will be found, on trial, that the same impulse given to a mass of two pounds will move it half an inch; and hence it is inferred, that, if applied to a mass weighing a million of pounds, it would move it the millionth part of an inch.

It is one of the curious results of the second law of motion, that an unlimited number of motions may exist together in the same body. Thus, at the same moment, we may be walking around a post in the cabin of a steam-boat, accompanying the boat in its passage around an island, revolving with the earth on its axis, flying through space in our annual circuit around the sun, and possibly wheeling, along with the sun and his whole retinue of planets, around some centre in common with the starry worlds.

The THIRD LAW of motion is this: *action and reaction are equal, and in contrary directions.*

Whenever I give a blow, the body struck exerts an equal force on the striking body. If I strike the water with an oar, the water communicates an equal impulse to the oar, which, being communicated to the boat, drives it forward in the opposite direction. If a magnet attracts a piece of iron, the iron attracts the magnet just as much, in the opposite direction; and, in short, every portion of matter in the universe attracts and is attracted by every other, equally, in an opposite direction. This brings us to the doctrine of universal gravitation, which is the very key that unlocks all the secrets of the skies. This will form the subject of my next Letter.

LETTER XIII.

TERRESTRIAL GRAVITY.

"To Him no high, no low, no great, no small,
He fills, He bounds, connects, and equals all."—*Pope*.

WE discover in Nature a tendency of every portion of matter towards every other. This tendency is called *gravitation*. In obedience to this power, a stone falls to the ground, and a planet revolves around the sun. We may contemplate this subject as it relates either to phenomena that take place near the surface of the earth, or in the celestial regions. The former, *gravity*, is exemplified by falling bodies; the latter, *universal gravitation*, by the motions of the heavenly bodies. The laws of terrestrial gravity were first investigated by Galileo; those of universal gravitation, by Sir Isaac Newton. Terrestrial gravity is only an individual example of universal gravitation; being the tendency of bodies towards the centre of the earth. We are so much accustomed, from our earliest years, to see bodies fall to the earth, that we imagine bodies must of necessity fall "downwards;" but when we reflect that the earth is round, and that bodies fall towards the centre on all sides of it, and that of course bodies on opposite sides of the earth fall in precisely opposite directions, and towards each other, we perceive that there must be some force acting to produce this effect; nor is it enough to say, as the ancients did, that bodies "naturally" fall to the earth. Every motion implies some force which produces it; and the fact that bodies fall towards the earth, on all sides of it, leads us to infer that that force, whatever it is, resides in the earth itself. We therefore call it *attraction*. We do not, however, say what attraction *is*, but what it *does*. We must bear in mind, also, that, according to the third law of motion, this attraction is mutual; that when a stone falls towards the earth, it exerts the same force on the earth that the earth exerts on the stone; but the motion of the earth towards the stone is as much less than that of the stone towards the earth, as its quantity of matter is greater; and therefore its motion is quite insensible.

But although we are compelled to acknowledge the *existence* of such a force as gravity, causing a tendency in all bodies towards each other, yet we know nothing of its *nature*, nor can we conceive by what medium bodies at such a distance as the moon and the earth exercise this influence on each other. Still, we may trace the modes in which this force acts; that is, its *laws*; for the laws of Nature are nothing else than the modes in which the powers of Nature act.

We owe chiefly to the great Galileo the first investigation of the laws of terrestrial gravity, as exemplified in falling bodies; and I will avail myself of this opportunity to make you better acquainted with one of the most interesting of men and greatest of philosophers.

Galileo was born at Pisa, in Italy, in the year 1564. He was the son of a Florentine nobleman, and was destined by his father for the medical profession, and to this his earlier studies were devoted. But a fondness and a genius for mechanical inventions had developed itself, at a very early age, in the construction of his toys, and a love of drawing; and as he grew older, a passion for mathematics, and for experimental research, predominated over his zeal for the study of medicine, and he fortunately abandoned that for the more congenial pursuits of natural philosophy and astronomy. In the twenty-fifth year of his age, he was appointed, by the Grand Duke of Tuscany, professor of mathematics in the University of Pisa. At that period, there prevailed in all the schools a most extraordinary reverence for the writings of Aristotle, the preceptor of Alexander the Great,—a philosopher who flourished in Greece, about three hundred years before the Christian era. Aristotle, by his great genius and learning, gained a wonderful ascendency over the minds of men, and became the oracle of the whole reading world for twenty centuries. It was held, on the one hand, that all truths worth knowing were contained in the writings of Aristotle; and, on the other, that an assertion which contradicted any thing in Aristotle could not be true. But Galileo had a greatness of mind which soared above the prejudices of the age in which he lived, and dared to interrogate Nature by the two great and only successful methods of discovering her secrets,— experiment and observation. Galileo was indeed the first philosopher that ever fully employed experiments as the means of learning the laws of Nature, by imitating on a small what she performs on a great scale, and thus detecting her modes of operation. Archimedes, the great Sicilian philosopher, had in ancient times introduced mathematical or geometrical reasoning into natural philosophy; but it was reserved for Galileo to unite the advantages of both mathematical and experimental reasonings in the study of Nature,—both sure and the only sure guides to truth, in this department of knowledge, at least. Experiment and observation furnish materials upon which geometry builds her reasonings, and from which she derives many truths that either lie for ever hidden from the eye of observation, or which it would require ages to unfold.

This method, of interrogating Nature by experiment and observation, was matured into a system by Lord Bacon, a celebrated English philosopher, early in the seventeenth century,—indeed, during the life of Galileo. Previous to that time, the inquirers into Nature did not open their eyes to see how the facts really *are*; but, by metaphysical processes, in imitation of Aristotle, determined how they *ought to be*, and hastily concluded that they were so. Thus, they did not study into the laws of motion, by observing how motion actually takes place, under various circumstances, but first, in their closets, constructed a definition of motion, and thence inferred all its properties. The system of reasoning respecting the phenomena of Nature, introduced by Lord Bacon, was this: in the first place, to examine all the facts of the case, and then from these to determine the laws of Nature. To derive general conclusions from the comparison of a great number of individual instances constitutes the peculiarity of the Baconian philosophy. It is called the *inductive* system, because its conclusions were built on the induction, or comparison, of a great many single facts. Previous to the time of Lord Bacon, hardly any insight had been gained into the causes of natural phenomena, and hardly one of the laws of Nature had been clearly established, because all the inquirers into Nature were upon a wrong road, groping their way through the labyrinth of error. Bacon pointed out to them the true path, and held before them the torch-light of experiment and observation, under whose guidance all successful students of Nature have since walked, and by whose illumination they have gained so wonderful an insight into the mysteries of the natural world.

It is a remarkable fact, that two such characters as Bacon and Galileo should appear on the stage at the same time, who, without any communication with each other, or perhaps without any personal knowledge of each other's existence, should have each developed the true method of investigating the laws of Nature. Galileo practised what Bacon only taught; and some, therefore, with much reason, consider Galileo as a greater philosopher than Bacon. "Bacon," says Hume, "pointed out, at a great distance, the road to philosophy; Galileo both pointed it out to others, and made, himself, considerable advances in it. The Englishman was ignorant of geometry; the Florentine revived that science, excelled in it, and was the first who applied it, together with experiment, to natural philosophy. The former rejected, with the most positive disdain, the system of Copernicus; the latter fortified it with new proofs, derived both from reason and the senses."

When we reflect that geometry is a science built upon self-evident truths, and that all its conclusions are the result of pure demonstration, and can admit of no controversy; when we further reflect, that experimental evidence rests on the testimony of the senses, and we infer a thing to be true because we actually see it to be so; it shows us the extreme bigotry, the darkness visible, that beclouded the human intellect, when it not only refused to admit conclusions first established by pure geometrical reasoning, and afterwards confirmed by experiments exhibited in the light of day, but instituted the most cruel persecutions against the great philosopher who first proclaimed these truths. Galileo was hated and persecuted by two distinct bodies of men, both possessing great influence in their respective spheres,—the one consisting of the learned doctors of philosophy, who did nothing more, from age to age, than reiterate the doctrines of Aristotle, and were consequently alarmed at the promulgation of principles subversive of those doctrines; the other consisting of the Romish priesthood, comprising the terrible Inquisition, who denounced the truths taught by Galileo, as inconsistent with certain declarations of the Holy Scriptures. We shall see, as we advance, what a fearful warfare he had to wage against these combined powers of darkness.

Aristotle had asserted, that, if two different weights of the same material were let fall from the same height, the heavier one would reach the ground sooner than the other, in proportion as it was more weighty. For example: if a ten-pound leaden weight and a one-pound were let fall from a given height at the same instant, the former would reach the ground ten times as soon as the latter. No one thought of making the trial, but it was deemed sufficient that Aristotle had said so; and accordingly this assertion had long been received as an axiom in the science of motion. Galileo ventured to appeal from the authority of Aristotle to that of his own senses, and maintained, that both weights would fall in the same time. The learned doctors ridiculed the idea. Galileo tried the experiment in their presence, by letting fall, at the same instant, large and small weights from the top of the celebrated leaning tower of Pisa. Yet, with the sound of the two weights clicking upon the pavement at the same moment, they still maintained that the ten-pound weight would reach the ground in one tenth part of the time of the other, because they could quote the chapter and verse of Aristotle where the fact was asserted. Wearied and disgusted with the malice and folly of these Aristotelian philosophers, Galileo, at the age of twenty-eight, resigned his situation in the university of Pisa, and removed to Padua, in the university of which place he was elected professor of mathematics. Up to this period, Galileo had devoted himself chiefly to the studies of the laws of motion, and the other branches of mechanical philosophy. Soon afterwards, he began to publish his writings, in rapid succession, and became at once among the most conspicuous of his age,—a rank which he afterwards well sustained and greatly exalted, by the invention of the telescope, and by his numerous astronomical discoveries. I will reserve an account of these great achievements until we come to that part of astronomy to which they were more immediately related, and proceed, now, to explain to you the leading principles of *terrestrial gravity*, as exemplified in falling bodies.

First, *all bodies near the earth's surface fall in straight lines towards the centre of the earth.* We are not to infer from this fact, that there resides at the centre any peculiar force, as a great loadstone, for example, which attracts bodies towards itself; but bodies fall towards the centre of the sphere, because the combined attractions of all the particles of matter in the earth, each exerting its proper force upon the body, would carry it towards the centre. This may be easily illustrated by a diagram. Let B, Fig. 29, page 140, be the centre of the earth, and A a body without it. Every portion of matter in the earth exerts some force on A, to draw it down to the earth. But since there is just as much matter on one side of the line A B, as on the other side, each half exerts an equal force to draw the body towards itself; therefore it falls in the direction of the diagonal between the two forces. Thus, if we compare the effects of any two particles of matter at equal distances from the line A B, but on opposite sides of it, as *a, b,* while the force of the particle at *a* would tend to draw A in the direction of A *a,* that of *b* would draw it in the direction of A *b,* and it would fall in the line A B, half way between the two. The same would hold true of any other two corresponding particles of matter on different sides of the earth, in respect to a body situated in any place without it.

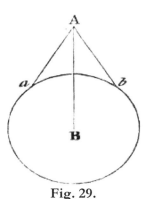

Fig. 29.

Secondly, *all bodies fall towards the earth, from the same height, with equal velocities.* A musket-ball, and the finest particle of down, if let fall from a certain height towards the earth, tend to descend towards it at the same rate, and would proceed with equal speed, were it not for the resistance of the air, which retards the down more than it does the ball, and finally stops it. If, however, the air be removed out of the way, as it may be by means of the air-pump, the two bodies keep side by side in falling from the greatest height at which we can try the experiment.

Thirdly, *bodies, in falling towards the earth, have their rate of motion continually accelerated.* Suppose we let fall a musket-ball from the top of a high tower, and watch its progress, disregarding the resistance of the air: the first second, it will pass over sixteen feet and one inch, but its speed will be constantly increased, being all the while urged onward by the same force, and retaining all that it has already acquired; so that the longer it is in falling, the swifter its motion becomes. Consequently, when bodies fall from a great height, they acquire an immense velocity before they reach the earth. Thus, a man falling from a balloon, or from the mast-head of a ship, is broken in pieces; and those meteoric stones, which sometimes fall from the sky, bury themselves deep in the earth. On measuring the spaces through which a body falls, it is found, that it will fall four times as far in two seconds as in one, and one hundred times as far in ten seconds as in one; and universally, the space described by a falling body is proportioned to the time multiplied into itself; that is, to the square of the time.

Fourthly, *gravity is proportioned to the quantity of matter.* A body which has twice as much matter as another exerts a force of attraction twice as great, and also receives twice as much from the same body as it would do, if it were only just as heavy as that body. Thus the earth, containing, as it does, forty times as much matter as the moon, exerts upon the moon forty times as much force as it would do, were its mass the same with that of the moon; but it is also capable of *receiving* forty times as much gravity from the moon as it would do, were its mass the same as the moon's; so that the power of attracting and that of being attracted are reciprocal; and it is therefore correct to say, that the moon attracts the earth *just as much* as the earth attracts the moon; and the same may be said of any two bodies, however different in quantity of matter.

Fifthly, *gravity, when acting at a distance from the earth, is not as intense as it is near the earth.* At such a distance as we are accustomed to ascend above the general level of the earth, no great difference is observed. On the tops of high mountains, we find bodies falling towards the earth, with nearly the same speed as they do from the smallest elevations. It is found, nevertheless, that there is a real difference; so that, in fact, the weight of a body (which is nothing more than the measure of its force of gravity) is not quite so great on the tops of high mountains as at the general level of the sea. Thus, a thousand pounds' weight, on the top of a mountain half a mile high, would weigh a quarter of a pound less than at the level of the sea; and if elevated four thousand miles above the earth,—that is, *twice* as far from the centre of the earth as the surface is from the centre,—it would weigh only one fourth as much as before; if *three times* as far, it would weigh only one ninth as much. So that the force of gravity decreases, as we recede from the earth, in the same proportion as the square of the distance increases. This fact is generalized by saying, that *the force of gravity, at different distances from the earth, is inversely as the square of the distance.*

Were a body to fall from a great distance,—suppose a thousand times that of the radius of the earth,—the force of gravity being one million times less than that at the surface of the earth, the motion of the body would be exceedingly slow, carrying it over only the sixth part of an inch in a day. It would be a long time, therefore, in making any sensible approaches towards the earth; but at length, as it drew near to the earth it would acquire a very great velocity, and would finally rush towards it with prodigious violence. Falling so far, and being continually accelerated on the way, we might suppose that it would at length attain a velocity infinitely great; but it can be demonstrated, that, if a body were to fall from an infinite distance, attracted to the earth only by gravity, it could never acquire a velocity greater than about seven miles per second. This, however, is a speed inconceivably great, being about eighteen times the greatest velocity that can be given to a cannon-ball, and more than twenty-five thousand miles per hour.

But the phenomena of falling bodies must have long been observed, and their laws had been fully investigated by Galileo and others, before the cause of their falling was understood, or any such principle as gravity, inherent in the earth and in all bodies, was applied to them. The developement of this great principle was the work of Sir Isaac Newton; and I will give you, in my next Letter, some particulars respecting the life and discoveries of this wonderful man.

LETTER XIV.

SIR ISAAC NEWTON.—UNIVERSAL GRAVITATION.—FIGURE OF

THE EARTH'S ORBIT.—PRECESSION OF THE EQUINOXES.

"The heavens are all his own; from the wild rule
Of whirling vortices, and circling spheres,
To their first great simplicity restored.
The schools astonished stood; but found it vain
To combat long with demonstration clear,
And, unawakened, dream beneath the blaze
Of truth. At once their pleasing visions fled,
With the light shadows of the morning mixed,
When Newton rose, our philosophic sun."—*Thomson's Elegy.*

SIR ISAAC NEWTON was born in Lincolnshire, England, in 1642, just one year after the death of Galileo. His father died before he was born, and he was a helpless infant, of a diminutive size, and so feeble a frame, that his attendants hardly expected his life for a single hour. The family dwelling was of humble architecture, situated in a retired but beautiful valley, and was surrounded by a small farm, which afforded but a scanty living to the widowed mother and her precious charge. The cut on page 144, Fig 30, represents the modest mansion, and the emblems of rustic life that first met the eyes of this pride of the British nation, and ornament of human nature. It will probably be found, that genius has oftener emanated from the cottage than from the palace.

Fig. 30.

The boyhood of Newton was distinguished chiefly for his ingenious mechanical contrivances. Among other pieces of mechanism, he constructed a windmill so curious and complete in its workmanship, as to excite universal admiration. After carrying it a while by the force of the wind, he resolved to substitute animal power, and for this purpose he inclosed in it a mouse, which he called the miller, and which kept the mill a-going by acting on a tread-wheel. The power of the mouse was brought into action by unavailing attempts to reach a portion of corn placed above the wheel. A water-clock, a four-wheeled carriage propelled by the rider himself, and kites of superior workmanship, were among the productions of the mechanical genius of this gifted boy. At a little later period, he began to turn his attention to the motions of the heavenly bodies, and constructed several sun-dials on the walls of the house where he lived. All this was before he had reached his fifteenth year. At this age, he was sent by his mother, in company with an old family servant, to a neighboring market-town, to dispose of products of their farm, and to buy articles of merchandise for their family use; but the young philosopher left all these negotiations to his worthy partner, occupying himself, mean-while, with a collection of old books, which he had found in a garret. At other times, he stopped on the road, and took shelter with his book under a hedge, until the servant returned. They endeavored to educate him as a farmer; but the perusal of a book, the construction of a water-mill, or some other mechanical or scientific amusement, absorbed all his thoughts, when the sheep were going astray, and the cattle were devouring or treading down the corn. One of his uncles having found him one day under a hedge, with a book in his hand, and entirely absorbed in meditation, took it from him, and found that it was a mathematical problem which so engrossed his attention. His friends, therefore, wisely resolved to favor the bent of his genius, and removed him from the farm to the school, to prepare for the university. In the eighteenth year of his age, Newton was admitted into Trinity College, Cambridge. He made rapid and extraordinary advances in the mathematics, and soon afforded unequivocal presages of that greatness which afterwards placed him at the head of the human intellect. In 1669, at the age of twenty-seven, he became professor of mathematics at Cambridge, a post which he occupied for many years afterwards. During the four or five years previous to this he had, in fact, made most of those great discoveries which have immortalized his name. We are at present chiefly interested in one of these, namely, that of *universal gravitation*; and let us see by what steps he was conducted to this greatest of scientific discoveries.

In the year 1666, when Newton was about twenty-four years of age, the plague was prevailing at Cambridge, and he retired into the country. One day, while he sat in a garden, musing on the phenomena of Nature around him, an apple chanced to fall to the ground. Reflecting on the mysterious power that makes all bodies near the earth fall towards its centre, and considering that this power remains unimpaired at considerable heights above the earth, as on the tops of trees and mountains, he asked himself,— "May not the same force extend its influence to a great distance from the earth, even as far as the moon? Indeed, may not this be the very reason, why the moon is drawn away continually from the straight line in which every body tends to move, and is thus made to circulate around the earth?" You will recollect that it was mentioned, in my Letter which contained an account of the first law of motion, that if a body is put in motion by any force, it will always move forward in a straight line, unless some other force compels it to turn aside from such a direction; and that, when we see a body moving in a curve, as a circular orbit, we are authorized to conclude that there is some force existing within the circle, which continually draws the body away from the direction in which it tends to move. Accordingly, it was a very natural suggestion, to one so well acquainted with the laws of motion as Newton, that the moon should constantly bend towards the earth, from a tendency to fall towards it, as any other heavy body would do, if carried to such a distance from the earth. Newton had already proved, that if such a power as gravity extends from the earth to distant bodies, it must decrease, as the square of the distance from the centre of the earth increases; that is, at double the distance, it would be four times less; at ten times the distance, one hundred times less; and so on. Now, it was known that the moon is about sixty times as far from the centre of the earth as the surface of the earth is from the centre, and consequently, the force of attraction at the moon must be the square of sixty, or thirty-six hundred times less than it is at the earth; so that a body at the distance of the moon would fall towards the earth very slowly, only one thirty-six hundredth part as far in a given time, as at the earth. Does the moon actually fall towards the earth at this rate; or, what is the same thing, does she depart at this rate continually from the straight line in which she tends to move, and in which she would move, if no external force diverted her from it? On making the calculation, such was found to be the fact. Hence gravity, and no other force than gravity, acts upon the moon, and compels her to revolve around the earth. By reasonings equally conclusive, it was afterwards proved, that a similar force compels all the planets to circulate around the sun; and now, we may ascend from the contemplation of this force, as we have seen it exemplified in falling bodies, to that of a universal power whose influence extends to all the material creation. It is in this sense that we recognise the principle of universal gravitation, the law of which may be thus enunciated; *all bodies in the universe, whether great or small, attract each other, with forces proportioned to their respective quantities of matter, and inversely as the squares of their distances from each other.*

This law asserts, first, that attraction reigns throughout the material world, affecting alike the smallest particle of matter and the greatest body; secondly, that it acts upon every mass of matter, precisely in proportion to its quantity; and, thirdly, that its intensity is diminished as the square of the distance is increased.

Observation has fully confirmed the prevalence of this law throughout the solar system; and recent discoveries among the fixed stars, to be more fully detailed hereafter, indicate that the same law prevails there. The law of universal gravitation is therefore held to be the grand principle which governs all the celestial motions. Not only is it consistent with all the observed motions of the heavenly bodies, even the most irregular of those motions, but, when followed out into all its consequences, it would be competent to assert that such irregularities must take place, even if they had never been observed.

Newton first published the doctrine of universal gravitation in the 'Principia,' in 1687. The name implies that the work contains the fundamental principles of natural philosophy and astronomy. Being founded upon the immutable basis of mathematics, its conclusions must of course be true and unalterable, and thenceforth we may regard the great laws of the universe as traced to their remotest principle. The greatest astronomers and mathematicians have since occupied themselves in following out the plan which Newton began, by applying the principles of universal gravitation to all the subordinate as well as to the grand movements of the spheres. This great labor has been especially achieved by La Place, a French mathematician of the highest eminence, in his profound work, the 'Mecanique Celeste.' Of this work, our distinguished countryman, Dr. Bowditch, has given a magnificent translation, and accompanied it with a commentary, which both illustrates the original, and adds a great amount of matter hardly less profound than that.

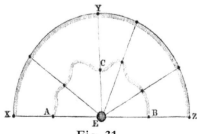

Fig. 31.

We have thus far taken the earth's orbit around the sun as a great circle, such being its projection on the sphere constituting the celestial ecliptic. The real path of the earth around the sun is learned, as I before explained to you, by the apparent path of the sun around the earth once a year. Now, when a body revolves about the earth at a great distance from us, as is the case with the sun and moon, we cannot certainly infer that it moves in a circle because it appears to describe a circle on the face of the sky, for such might be the appearance of its orbit, were it ever so irregular a curve. Thus, if E, Fig. 31, represents the earth, and ACB, the irregular path of a body revolving about it, since we should refer the body continually to some place on the celestial sphere, XYZ, determined by lines drawn from the eye to the concave sphere through the body, the body, while moving from A to B through C, would appear to move from X to Z, through Y. Hence, we must determine from other circumstances than the actual appearance, what is the true figure of the orbit.

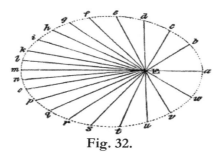

Fig. 32.

Were the earth's path a circle, having the sun in the centre, the sun would always appear to be at the same distance from us; that is, the radius of the orbit, or *radius vector,* (the name given to a line drawn from the centre of the sun to the orbit of any planet,) would always be of the same length. But the earth's distance from the sun is constantly varying, which shows that its orbit is not a circle. We learn the true figure of the orbit, by ascertaining the *relative distances* of the earth from the sun, at various periods of the year. These distances all being laid down in a diagram, according to their respective lengths, the extremities, on being connected, give us our first idea of the shape of the orbit, which appears of an oval form, and at least resembles an ellipse; and, on further trial, we find that it has the properties of an ellipse. Thus, let E, Fig. 32, be the place of the earth, and *a, b, c,* &c., successive positions of the sun; the *relative* lengths of the lines E *a,* E *b,* &c., being known, on connecting the points *a, b, c,* &c., the resulting figure indicates the true figure of the earth's orbit.

These relative distances are found in two different ways; first, *by changes in the sun's apparent diameter,* and, secondly, *by variations in his angular velocity.*

The same object appears to us smaller in proportion as it is more distant; and if we see a heavenly body varying in size, at different times, we infer that it is at different distances from us; that when largest, it is nearest to us, and when smallest, furthest off. Now, when the sun's diameter is accurately measured by instruments, it is found to vary from day to day; being, when greatest, more than thirty-two minutes and a half, and when smallest, only thirty-one minutes and a half,—differing, in all, about seventy-five seconds. When the diameter is greatest, which happens in January, we know that the sun is nearest to us; and when the diameter is least, which occurs in July, we infer that the sun is at the greatest distance from us. The point where the earth, or any planet, in its revolution, is nearest the sun, is called its *perihelion*; the point where it is furthest from the sun, its *aphelion*. Suppose, then, that, about the first of January, when the diameter of the sun is greatest, we draw a line, E *a*, Fig. 32, to represent it, and afterwards, every ten days, draw other lines, E *b*, E *c*, &c.; increasing in the same ratio as the apparent diameters of the sun decrease. These lines must be drawn at such a distance from each other, that the triangles, E *a b*, E *b c*, &c., shall be all equal to each other, for a reason that will be explained hereafter. On connecting the extremities of these lines, we shall obtain the figure of the earth's orbit.

Similar conclusions may be drawn from observations on the sun's *angular velocity*. A body appears to move most rapidly when nearest to us. Indeed, the apparent velocity increases rapidly, as it approaches us, and as rapidly diminishes, when it recedes from us. If it comes twice as near as before, it appears to move not merely twice as swiftly, but four times as swiftly; if it comes ten times nearer, its apparent velocity is one hundred times as great as before. We say, therefore, that the velocity varies inversely as the square of the distance; for, as the distance is diminished ten times, the velocity is increased the square of ten; that is, one hundred times. Now, by noting the time it takes the sun, from day to day, to cross the central wire of the transit-instrument, we learn the comparative velocities with which it moves at different times; and from these we derive the comparative distances of the sun at the corresponding times; and laying down these relative distances in a diagram, as before, we get our first notions of the actual figure of the earth's orbit, or the path which it describes in its annual revolution around the sun.

Having now learned the fact, that the earth moves around the sun, not in a circular but in an elliptical orbit, you will desire to know by what forces it is impelled, to make it describe this figure, with such uniformity and constancy, from age to age. It is commonly said, that gravity causes the earth and the planets to circulate around the sun; and it is true that it is gravity which turns them aside from the straight line in which, by the first law of motion, they tend to move, and thus causes them to revolve around the sun. But what force is that which gave to them this original impulse, and impressed upon them such a tendency to move forward in a straight line? The name *projectile* force is given to it, because it is the same *as though* the earth were originally projected into space, when first created; and therefore its motion is the result of two forces, the projectile force, which would cause it to move forward in a straight line which is a tangent to its orbit, and gravitation, which bends it towards the sun. But before you can clearly understand the nature of this motion, and the action of the two forces that produce it, I must explain to you a few elementary principles upon which this and all the other planetary motions depend.

You have already learned, that when a body is acted on by two forces, in different directions, it moves in the direction of neither, but in some direction between them. If I throw a stone horizontally, the attraction of the earth will continually draw it downward, out of the line of direction in which it was thrown, and make it descend to the earth in a curve. The particular form of the curve will depend on the velocity with which it is thrown. It will always *begin* to move in the line of direction in which it is projected; but it will soon be turned from that line towards the earth. It will, however, continue nearer to the line of projection in proportion as the velocity of projection is greater. Thus, let A C, Fig. 33, be perpendicular to the horizon, and A B parallel to it, and let a stone be thrown from A, in the direction of A B. It will, in every case, commence its motion in the line A B, which will therefore be a tangent to the curve it describes; but, if it is thrown with a small velocity, it will soon depart from the tangent, describing the line A D; with a greater velocity, it will describe a curve nearer the tangent, as A E; and with a still greater velocity, it will describe the curve A F.

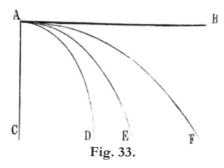

Fig. 33.

As an example of a body revolving in an orbit under the influence of two forces, suppose a body placed at any point, P, Fig. 34, above the surface of the earth, and let P A be the direction of the earth's centre; that is, a line perpendicular to the horizon. If the body were allowed to move, without receiving any impulse, it would descend to the earth in the direction P A with an accelerated motion. But suppose that, at the moment of its departure from P, it receives a blow in the direction P B, which would carry it to B in the time the body would fall from P to A; then, under the influence of both forces, it would descend along the curve P D. If a stronger blow were given to it in the direction P B, it would describe a larger curve, P E; or, finally, if the impulse were sufficiently strong, it would circulate quite around the earth, and return again to P, describing the circle P F G. With a velocity of projection still greater, it would describe an ellipse, P I K; and if the velocity be increased to a certain degree, the figure becomes a parabola, L P M,—a curve which never returns into itself.

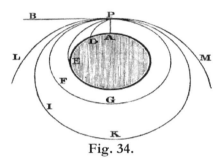

Fig. 34.

In Fig. 35, page 154, suppose the planet to have passed the point C, at the aphelion, with so small a velocity, that the attraction of the sun bends its path very much, and causes it immediately to begin to approach towards the sun. The sun's attraction will increase its velocity, as it moves through D, E, and F, for the sun's attractive force on the planet, when at D, is acting in the direction D S; and, on account of the small angle made between D E and D S, the force acting in the line D S helps the planet forward in the path D E, and thus increases its velocity. In like manner, the velocity of the planet will be continually increasing as it passes through D, E, and F; and though the attractive force, on account of the planet's nearness, is so much increased, and tends, therefore, to make the orbit more curved, yet the velocity is also so much increased, that the orbit is not more curved than before; for the same increase of velocity, occasioned by the planet's approach to the sun, produces a greater increase of centrifugal force, which carries it off again. We may see, also, the reason why, when the planet has reached the most distant parts of its orbit, it does not entirely fly off, and never return to the sun; for, when the planet passes along H, K, A, the sun's attraction retards the planet, just as gravity retards a ball rolled up hill; and when it has reached C, its velocity is very small, and the attraction to the centre of force causes a great deflection from the tangent, sufficient to give its orbit a great curvature, and the planet wheels about, returns to the sun, and goes over the same orbit again. As the planet recedes from the sun, its centrifugal force diminishes faster than the force of gravity, so that the latter finally preponderates.

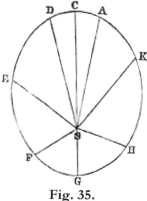

Fig. 35.

I shall conclude what I have to say at present, respecting the motion of the earth around the sun, by adding a few words respecting the precession of the equinoxes.

The *precession of the equinoxes* is a slow but continual shifting of the equinoctial points, from east to west. Suppose that we mark the exact place in the heavens where, during the present year, the sun crosses the equator, and that this point is close to a certain star; next year, the sun will cross the equator a little way westward of that star, and so every year, a little further westward, until, in a long course of ages, the place of the equinox will occupy successively every part of the ecliptic, until we come round to the same star again. As, therefore, the sun revolving from west to east, in his apparent orbit, comes round to the point where it left the equinox, it meets the equinox before it reaches that point. The appearance is as though the equinox *goes forward* to meet the sun, and hence the phenomenon is called the *precession* of the equinoxes; and the fact is expressed by saying, that the equinoxes retrograde on the ecliptic, until the line of the equinoxes (a straight line drawn from one equinox to the other) makes a complete revolution, from east to west. This is of course a retrograde motion, since it is contrary to the order of the signs. The equator is conceived as *sliding* westward on the ecliptic, always preserving the same inclination to it, as a ring, placed at a small angle with another of nearly the same size which remains fixed, may be slid quite around it, giving a corresponding motion to the two points of intersection. It must be observed, however, that this mode of conceiving of the precession of the equinoxes is purely imaginary, and is employed merely for the convenience of representation.

The amount of precession annually is fifty seconds and one tenth; whence, since there are thirty-six hundred seconds in a degree, and three hundred and sixty degrees in the whole circumference of the ecliptic, and consequently one million two hundred and ninety-six thousand seconds, this sum, divided by fifty seconds and one tenth, gives twenty-five thousand eight hundred and sixty-eight years for the period of a complete revolution of the equinoxes.

Suppose we now fix to the centre of each of the two rings, before mentioned, a wire representing its axis, one corresponding to the axis of the ecliptic, the other to that of the equator, the extremity of each being the pole of its circle. As the ring denoting the equator turns round on the ecliptic, which, with its axis, remains fixed, it is easy to conceive that the axis of the equator revolves around that of the ecliptic, and the pole of the equator around the pole of the ecliptic, and constantly at a distance equal to the inclination of the two circles. To transfer our conceptions to the celestial sphere, we may easily see that the axis of the diurnal sphere (that of the earth produced) would not have its pole constantly in the same place among the stars, but that this pole would perform a slow revolution around the pole of the ecliptic, from east to west, completing the circuit in about twenty-six thousand years. Hence the star which we now call the pole-star has not always enjoyed that distinction, nor will it always enjoy it, hereafter. When the earliest catalogues of the stars were made, this star was twelve degrees from the pole. It is now one degree twenty-four minutes, and will approach still nearer; or, to speak more accurately, the pole will come still nearer to this star, after which it will leave it, and successively pass by others. In about thirteen thousand years, the bright star Lyra (which lies near the circle in which the pole of the equator revolves about the pole of the ecliptic, on the side opposite to the present pole-star) will be within five degrees of the pole, and will constitute the pole-star. As Lyra now passes near our zenith, you might suppose that the change of position of the pole among the stars would be attended with a change of altitude of the north pole above the horizon. This mistaken idea is one of the many misapprehensions which result from the habit of considering the horizon as a fixed circle in space. However the pole might shift its position in space, we should still be at the same distance from it, and our horizon would always reach the same distance beyond it.

The time occupied by the sun, in passing from the equinoctial point round to the same point again, is called the *tropical year*. As the sun does not perform a complete revolution in this interval, but falls short of it fifty seconds and one tenth, the tropical year is shorter than the sidereal by twenty minutes and twenty seconds, in mean solar time, this being the time of describing an arc of fifty seconds and one tenth, in the annual revolution.

The changes produced by the precession of the equinoxes, in the apparent places of the circumpolar stars, have led to some interesting results in *chronology*. In consequence of the retrograde motion of the equinoctial points, the *signs* of the ecliptic do not correspond, at present, to the *constellations* which bear the same names, but lie about one sign, or thirty degrees, westward of them. Thus, that division of the ecliptic which is called the sign Taurus lies in the constellation Aries, and the sign Gemini, in the constellation Taurus. Undoubtedly, however, when the ecliptic was thus first divided, and the divisions named, the several constellations lay in the respective divisions which bear their names.

LETTER XV.

THE MOON.

"Soon as the evening shades prevail
The Moon takes up the wondrous tale,
And nightly to the listening earth
Repeats the story of her birth."—*Addison.*

HAVING now learned so much of astronomy as relates to the earth and the sun, and the mutual relations which exist between them, you are prepared to enter with advantage upon the survey of the other bodies that compose the solar system. This being done, we shall then have still before us the boundless range of the fixed stars.

The moon, which next claims our notice, has been studied by astronomers with greater attention than any other of the heavenly bodies, since her comparative nearness to the earth brings her peculiarly within the range of our telescopes, and her periodical changes and very irregular motions, afford curious subjects, both for observation and speculation. The mild light of the moon also invites our gaze, while her varying aspects serve barbarous tribes, especially, for a kind of dial-plate inscribed on the face of the sky, for weeks, and months, and times, and seasons.

The moon is distant from the earth about two hundred and forty thousand miles; or, more exactly, two hundred and thirty-eight thousand five hundred and forty-five miles. Her angular or apparent diameter is about half a degree, and her real diameter, two thousand one hundred and sixty miles. She is a companion, or satellite, to the earth, revolving around it every month, and accompanying us in our annual revolution around the sun. Although her nearness to us makes her appear as a large and conspicuous object in the heavens, yet, in comparison with most of the other celestial bodies, she is in fact very small, being only one forty-ninth part as large as the earth, and only about one seventy millionth part as large as the sun.

The moon shines by light borrowed from the sun, being itself an opaque body, like the earth. When the disk, or any portion of it, is illuminated, we can plainly discern, even with the naked eye, varieties of light and shade, indicating inequalities of surface which we imagine to be land and water. I believe it is the common impression, that the darker portions are land and the lighter portions water; but if either part is water, it must be the darker regions. A smooth polished surface, like water, would reflect the sun's light like a mirror. It would, like a convex mirror, form a diminished image of the sun, but would not itself appear luminous like an uneven surface, which multiplies the light by numerous reflections within itself. Thus, from this cause, high broken mountainous districts appear more luminous than extensive plains.

Figures 36, 37. TELESCOPIC VIEWS OF THE MOON.

By the aid of the telescope, we may see undoubted indications of mountains and valleys. Indeed, with a good glass, we can discover the most decisive evidence that the surface of the moon is exceedingly varied,—one part ascending in lofty peaks, another clustering in huge mountain groups, or long ranges, and another bearing all the marks of deep caverns or valleys. You will not, indeed, at the first sight of the moon through a telescope, recognise all these different objects. If you look at the moon when half her disk is enlightened, (which is the best time for seeing her varieties of surface,) you will, at the first glance, observe a motley appearance, particularly along the line called the *terminator*, which separates the enlightened from the unenlightened part of the disk. (Fig. 37.) On one side of the terminator, within the dark part of the disk, you will see illuminated points, and short, crooked lines, like rude characters marked with chalk on a black ground. On the other side of the terminator you will see a succession of little circular groups, appearing like numerous bubbles of oil on the surface of water. The further you carry your eye from the terminator, on the same side of it, the more indistinctly formed these bubbles appear, until towards the edge of the moon they assume quite a different aspect.

Some persons, when they look into a telescope for the first time, having heard that mountains and valleys are to be seen, and discovering nothing but these unmeaning figures, break off in disappointment, and have their faith in these things rather diminished than increased. I would advise you, therefore, before you take even your first view of the moon through a telescope, to form as clear an idea as you can, how mountains, and valleys, and caverns, situated at such a distance from the eye, ought to look, and by what marks they may be recognised. Seize, if possible, the most favorable period, (about the time of the first quarter,) and previously learn from drawings and explanations, how to interpret every thing you see.

What, then, ought to be the respective appearances of mountains, valleys, and deep craters, or caverns, in the moon? The sun shines on the moon in the same way as it shines on the earth; and let, us reflect, then, upon the manner in which it strikes similar objects here. One half the globe is constantly enlightened; and, by the revolution of the earth on its axis, the terminator, or the line which separates the enlightened from the unenlightened part of the earth, travels along from east to west, over different places, as we see the moon's terminator travel over her disk from new to full moon; although, in the case of the earth, the motion is more rapid, and depends on a different cause. In the morning, the sun's light first strikes upon the tops of the mountains, and, if they are very high, they may be brightly illuminated while it is yet night in the valleys below. By degrees, as the sun rises, the circle of illumination travels down the mountain, until at length it reaches the bottom of the valleys; and these in turn enjoy the full light of day. Again, a mountain casts a shadow opposite to the sun, which is very long when the sun first rises, and shortens continually as the sun ascends, its length at a given time, however, being proportioned to the height of the mountain; so that, if the shadow be still very long when the sun is far above the horizon, we infer that the mountain is very lofty. We may, moreover, form some judgment of the shape of a mountain, by observing that of its shadow.

Now, the moon is so distant that we could not easily distinguish places simply by their elevations, since they would be projected into the same imaginary plane which constitutes the apparent disk of the moon; but the foregoing considerations would enable us to infer their existence. Thus, when you view the moon at any time within her first quarter, but better near the end of that period, you will observe, on the side of the terminator within the dark part of the disk, the tops of mountains which the light of the sun is just striking, as the morning sun strikes the tops of mountains on the earth. These you will recognise by those white specks and little crooked lines, before mentioned, as is represented in Fig. 37. These bright points and lines you will see altering their figure, every hour, as they come more and more into the sun's light; and, mean-while, other bright points, very minute at first, will start into view, which also in turn grow larger as the terminator approaches them, until they fall into the enlightened part of the disk. As they fall further and further within this part, you will have additional proofs that they are mountains, from the shadows which they cast on the plain, always in a direction opposite to the sun. The mountain itself may entirely disappear, or become confounded with the other enlightened portions of the surface; but its position and its shape may still be recognised by the dark line which it projects on the plane. This line will correspond in shape to that of the mountain, presenting at one time a long serpentine stripe of black, denoting that the mountain is a continued range; at another time exhibiting a conical figure tapering to a point, or a series of such sharp points; or a serrated, uneven termination, indicating, in each case respectively, a conical mountain, or a group of peaks, or a range with lofty cliffs. All these appearances will indeed be seen in miniature; but a little familiarity with them will enable you to give them, in imagination, their proper dimensions, as you give to the pictures of known animals their due sizes, although drawn on a scale far below that of real life.

In the next place, let us see how valleys and deep craters in the moon might be expected to appear. We could not expect to see depressions any more than elevations, since both would alike be projected on the same imaginary disk. But we may recognise such depressions, from the manner in which the light of the sun shines into them. When we hold a china tea-cup at some distance from a candle, in the night, the candle being elevated but little above the level of the top of the cup, a luminous crescent will be formed on the side of the cup opposite to the candle, while the side next to the candle will be covered by a deep shadow. As we gradually elevate the candle, the crescent enlarges and travels down the side of the cup, until finally the whole interior becomes illuminated. We observe similar appearances in the moon, which we recognise as deep depressions. They are those circular spots near the terminator before spoken of, which look like bubbles of oil floating on water. They are nothing else than circular craters or deep valleys. When they are so situated that the light of the sun is just beginning to shine into them, you may see, as in the tea-cup, a luminous crescent around the side furthest from the sun, while a deep black shadow is cast on the side next to the sun. As the cavity is turned more and more towards the light, the crescent enlarges, until at length the whole interior is illuminated. If the tea-cup be placed on a table, and a candle be held at some distance from it, nearly on a level with the top, but a little above it, the cup itself will cast a shadow on the table, like any other elevated object. In like manner, many of these circular spots on the moon cast deep shadows behind them, indicating that the tops of the craters are elevated far above the general level of the moon. The regularity of some of these circular spots is very remarkable. The circle, in some instances, appears as well formed as could be described by a pair of compasses, while in the centre there not unfrequently is seen a conical mountain casting its pointed shadow on the bottom of the crater. I hope you will enjoy repeated opportunities to view the moon through a telescope. Allow me to recommend to you, not to rest satisfied with a hasty or even with a single view, but to verify the preceding remarks by repeated and careful inspection of the lunar disk, at different ages of the moon.

The various places on the moon's disk have received appropriate names. The dusky regions being formerly supposed to be seas, were named accordingly; and other remarkable places have each two names, one derived from some well-known spot on the earth, and the other from some distinguished personage. Thus, the same bright spot on the surface of the moon is called *Mount Sinai* or *Tycho*, and another, *Mount Etna* or *Copernicus*. The names of individuals, however, are more used than the others. The diagram, Fig. 36, (see page 159,) represents rudely, the telescopic appearance of the full moon. The reality is far more beautiful. A few of the most remarkable points have the following names corresponding to the numbers and letters on the map.

1. Tycho,	6. Eratosthenes,
2. Kepler,	7. Plato,
3. Copernicus,	8. Archimedes,
4. Aristarchus,	9. Eudoxus,
5. Helicon,	10. Aristotle.

- A. Mare Humorum, *Sea of Humors*,
- B. Mare Nubium, *Sea of Clouds*,
- C. Mare Imbrium, *Sea of Rains*,
- D. Mare Nectaris, *Sea of Nectar*,
- E. Mare Tranquillitatis, *Sea of Tranquillity*,
- F. Mare Serenitatis, *Sea of Serenity*,
- G. Mare Fecunditatis, *Sea of Plenty*,
- H. Mare Crisium, *Crisian Sea*.

The heights of the lunar mountains, and the depths of the valleys, can be estimated with a considerable degree of accuracy. Some of the mountains are as high as five miles, and the valleys, in some instances, are four miles deep. Hence it is inferred, that the surface of the moon is more broken and irregular than that of the earth, its mountains being higher and its valleys deeper, in proportion to its magnitude, than those of the earth.

The varieties of surface in the moon, as seen by the aid of large telescopes, have been well described by Dr. Dick, in his 'Celestial Scenery,' and I cannot give you a better idea of them, than to add a few extracts from his work. The lunar mountains in general exhibit an arrangement and an aspect very different from the mountain scenery of our globe. They may be arranged under the four following varieties:

First, *insulated mountains*, which rise from plains nearly level, shaped like a sugar loaf, which may be supposed to present an appearance somewhat similar to Mount Etna, or the Peak of Teneriffe. The shadows of these mountains, in certain phases of the moon, are as distinctly perceived as the shadow of an upright staff, when placed opposite to the sun; and these heights can be calculated from the length of their shadows. Some of these mountains being elevated in the midst of extensive plains, would present to a spectator on their summits magnificent views of the surrounding regions.

Secondly, *mountain ranges*, extending in length two or three hundred miles. These ranges bear a distant resemblance to our Alps, Apennines, and Andes; but they are much less in extent. Some of them appear very rugged and precipitous; and the highest ranges are in some places more than four miles in perpendicular altitude. In some instances, they are nearly in a straight line from northeast to southwest, as in the range called the *Apennines*; in other cases, they assume the form of a semicircle, or crescent.

Thirdly, *circular ranges*, which appear on almost every part of the moon's surface, particularly in its southern regions. This is one grand peculiarity of the lunar ranges, to which we have nothing similar on the earth. A plain, and sometimes a large cavity, is surrounded with a circular ridge of mountains, which encompasses it like a mighty rampart. These annular ridges and plains are of all dimensions, from a mile to forty or fifty miles in diameter, and are to be seen in great numbers over every region of the moon's surface; they are most conspicuous, however, near the upper and lower limbs, about the time of the half moon.

The mountains which form these circular ridges are of different elevations, from one fifth of a mile to three miles and a half, and their shadows cover one half of the plain at the base. These plains are sometimes on a level with the general surface of the moon, and in other cases they are sunk a mile or more below the level of the ground which surrounds the exterior circle of the mountains.

Fourthly, *central mountains*, or those which are placed in the middle of circular plains. In many of the plains and cavities surrounded by circular ranges of mountains there stands a single insulated mountain, which rises from the centre of the plain, and whose shadow sometimes extends, in the form of a pyramid, half across the plain to the opposite ridges. These central mountains are generally from half a mile to a mile and a half in perpendicular altitude. In some instances, they have two, and sometimes three, different tops, whose shadows can be easily distinguished from each other. Sometimes they are situated towards one side of the plain, or cavity; but in the great majority of instances their position is nearly or exactly central. The lengths of their bases vary from five to about fifteen or sixteen miles.

The *lunar caverns* form a very peculiar and prominent feature of the moon's surface, and are to be seen throughout almost every region, but are most numerous in the southwest part of the moon. Nearly a hundred of them, great and small, may be distinguished in that quarter. They are all nearly of a circular shape, and appear like a very shallow egg-cup. The smaller cavities appear, within, almost like a hollow cone, with the sides tapering towards the centre; but the larger ones have, for the most part, flat bottoms, from the centre of which there frequently rises a small, steep, conical hill, which gives them a resemblance to the circular ridges and central mountains before described. In some instances, their margins are level with the general surface of the moon; but, in most cases, they are encircled with a high annular ridge of mountains, marked with lofty peaks. Some of the larger of these cavities contain smaller cavities of the same kind and form, particularly in their sides. The mountainous ridges which surround these cavities reflect the greatest quantity of light; and hence that region of the moon in which they abound appears brighter than any other. From their lying in every possible direction, they appear, at and near the time of full moon, like a number of brilliant streaks, or radiations. These radiations appear to converge towards a large brilliant spot, surrounded by a faint shade, near the lower part of the moon, which is named Tycho,—a spot easily distinguished even by a small telescope. The spots named Kepler and Copernicus are each composed of a central spot with luminous radiations.[8]

The broken surface and apparent geological structure of the moon has suggested the opinion, that the moon has been subject to powerful *volcanic* action. This opinion receives support from certain actual appearances of volcanic fires, which have at different times been observed. In a total eclipse of the sun, the moon comes directly between us and that luminary, and presents her dark side towards us under circumstances very favorable for observation. At such times, several astronomers, at different periods, have noticed bright spots, which they took to be volcanoes. It must evidently require a large fire to be visible at all, at such a distance; and even a burning spark, or point but just visible in a large telescope, might be in fact a volcano raging like Etna or Vesuvius. Still, as fires might be supposed to exist in the moon from different causes, we should require some marks peculiar to volcanic fires, to assure us that such was their origin in a given case. Dr. Herschel examined this point with great attention, and with better means of observation than any of his predecessors enjoyed, and fully embraced the opinion that what he saw were volcanoes. In April, 1787, he records his observations as follows: "I perceive three volcanoes in different places in the dark part of the moon. Two of them are already nearly extinct, or otherwise in a state of going to break out; the third shows an eruption of fire or luminous matter." On the next night, he says: "The volcano burns with greater violence than last night; its diameter cannot be less than three seconds; and hence the shining or burning matter must be above three miles in diameter. The appearance resembles a small piece of burning charcoal, when it is covered with a very thin coat of white ashes; and it has a degree of brightness about as strong as that with which such a coal would be seen to glow in faint daylight." That these were really volcanic fires, he considered further evident from the fact, that where a fire, supposed to have been volcanic, had been burning, there was seen, after its extinction, an accumulation of matter, such as would arise from the production of a great quantity of lava, sufficient to form a mountain.

It is probable that the moon has an *atmosphere*, although it is difficult to obtain perfectly satisfactory evidence of its existence; for granting the existence of an atmosphere bearing the same proportion to that planet as our atmosphere bears to the earth, its dimensions and its density would be so small, that we could detect its presence only by the most refined observations. As our twilight is owing to the agency of our atmosphere, so, could we discern any appearance of twilight in the moon, we should regard that fact as indicating that she is surrounded by an atmosphere. Or, when the moon covers the sun in a solar eclipse, could we see around her circumference a faint luminous ring, indicating that the sunlight shone through an aerial medium, we might likewise infer the existence of such a medium. Such a faint ring of light has sometimes, as is supposed, been observed. Schroeter, a German astronomer, distinguished for the acuteness of his vision and his powers of observation in general, was very confident of having obtained, from different sources, clear evidence of a lunar atmosphere. He concluded, that the inferior or more dense part of the moon's atmosphere is not more than fifteen hundred feet high, and that the entire height, at least to the limit where it would be too rare to produce any of the phenomena which are relied on as proofs of its existence, is not more than a mile.

It has been a question, much agitated among astronomers, whether there is *water* in the moon. Analogy strongly inclines us to reply in the affirmative. But the analogy between the earth and the moon, as derived from all the particulars in which we can compare the two bodies, is too feeble to warrant such a conclusion, and we must have recourse to other evidence, before we can decide the point. In the first place, then, there is no positive evidence in favor of the existence of water in the moon. Those extensive level regions, before spoken of, and denominated seas in the geography of this planet, have no other signs of being water, except that they are level and dark. But both these particulars would characterize an earthly plain, like the deserts of Arabia and Africa. In the second place, were those dark regions composed of water, the terminator would be entirely smooth where it passed over these oceans or seas. It is indeed indented by few inequalities, compared with those which it exhibits where it passes over the mountainous regions; but still, the inequalities are too considerable to permit the conclusion, that these level spots are such perfect levels as water would form. They do not appear to be more perfect levels than many plain countries on the globe. The deep caverns, moreover, seen in those dusky spots which were supposed to be seas, are unfavorable to the supposition that those regions are covered by water. In the third place, the face of the moon, when illuminated by the sun and not obscured by the state of our own atmosphere, is always serene, and therefore free from clouds. Clouds are objects of great extent; they frequently intercept light, like solid bodies; and did they exist about the moon, we should certainly see them, and should lose sight of certain parts of the lunar disk which they covered. But neither position is true; we neither see any clouds about the moon, with our best telescopes, nor do we, by the intervention of clouds, ever lose sight of any portion of the moon when our own atmosphere is clear. But the want of clouds in the lunar atmosphere almost necessarily implies the absence of water in the moon. This planet is at the same distance from the sun as our own, and has, in this respect, an equal opportunity to feel the influence of his rays. Its days are also twenty-seven times as long as ours, a circumstance which would augment the solar heat. When the pressure of the atmosphere is diminished on the surface of water, its tendency to pass into the state of vapor is increased. Were the whole pressure of the atmosphere removed from the surface of a lake, in a Summer's day, when the temperature was no higher than seventy-two degrees, the water would begin to boil. Now it is well ascertained, that if there be any atmosphere about the moon, it is much lighter than ours, and presses on the surface of that body with a proportionally small force. This circumstance, therefore, would conspire with the other causes mentioned, to convert all the water of the moon into vapor, if we could suppose it to have existed at any given time.

But those, who are anxious to furnish the moon and other planets with all the accommodations which they find in our own, have a subterfuge in readiness, to which they invariably resort in all cases like the foregoing. "There may be," say they, "some means, unknown to us, provided for retaining water on the surface of the moon, and for preventing its being wasted by evaporation: perhaps it remains unaltered in quantity, imparting to the lunar regions perpetual verdure and fertility." To this I reply, that the bare possibility of a thing is but slight evidence of its reality; nor is such a condition possible, except by miracle. If they grant that the laws of Nature are the same in the moon as in the earth, then, according to the foregoing reasoning, there cannot be water in the moon; but if they say that the laws of Nature are not the same there as here, then we cannot reason at all respecting them. One who resorts to a subterfuge of this kind ruins his own cause. He argues the existence of water in the moon, from the analogy of that planet to this. But if the laws of Nature are not the same there as here, what becomes of his analogy? A liquid substance which would not evaporate by such a degree of solar heat as falls on the moon, which would not evaporate the faster, in consequence of the diminished atmospheric pressure which prevails there, could not be water, for it would not have the properties of water, and things are known by their properties. Whenever we desert the cardinal principle of the Newtonian philosophy,—that the laws of Nature are uniform throughout all her realms,—we wander in a labyrinth; all analogies are made void; all physical reasonings cease; and imaginary possibilities or direct miracles take the place of legitimate natural causes.

On the supposition that the moon is inhabited, the question has often been raised, whether we may hope that our telescopes will ever be so much improved, and our other means of observation so much augmented, that we shall be able to discover either the lunar inhabitants or any of their works.

The improbability of our ever identifying *artificial structures* in the moon may be inferred from the fact, that a space a mile in diameter is the least space that could be distinctly seen. Extensive works of art, as large cities, or the clearing up of large tracts of country for settlement or tillage, might indeed afford some varieties of surface; but they would be merely varieties of light and shade, and the individual objects that occasioned them would probably never be recognised by their distinctive characters. Thus, a building equal to the great pyramid of Egypt, which covers a space less than the fifth of a mile in diameter, would not be distinguished by its figure; indeed, it would be a mere point. Still less is it probable that we shall ever discover any inhabitants in the moon. Were we to view the moon with a telescope that magnifies ten thousand times, it would bring the moon apparently ten thousand times nearer, and present it to the eye like a body twenty-four miles off. But even this is a distance too great for us to see the works of man with distinctness. Moreover, from the nature of the telescope itself, we can never hope to apply a magnifying power so high as that here supposed. As I explained to you, when speaking of the telescope, whenever we increase the magnifying power of this instrument we diminish its field of view, so that with very high magnifiers we can see nothing but a point, such as a fixed star. We at the same time, also, magnify the vapors and smoke of the atmosphere, and all the imperfections of the medium, which greatly obscures the object, and prevents our seeing it distinctly. Hence it is generally most satisfactory to view the moon with low powers, which afford a large field of view and give a clear light. With Clark's telescope, belonging to Yale College, we seldom gain any thing by applying to the moon a higher power than one hundred and eighty, although the instrument admits of magnifiers as high as four hundred and fifty.

Some writers, however, suppose that possibly we may trace indications of lunar inhabitants in their works, and that they may in like manner recognise the existence of the inhabitants of our planet. An author, who has reflected much on subjects of this kind, reasons as follows: "A navigator who approaches within a certain distance of a small island, although he perceives no human being upon it, can judge with certainty that it is inhabited, if he perceives human habitations, villages, corn-fields, or other traces of cultivation. In like manner, if we could perceive changes or operations in the moon, which could be traced to the agency of intelligent beings, we should then obtain satisfactory evidence that such beings exist on that planet; and it is thought possible that such operations may be traced. A telescope which magnifies twelve hundred times will enable us to perceive, as a visible point on the surface of the moon, an object whose diameter is only about three hundred feet. Such an object is not larger than many of our public edifices; and therefore, were any such edifices rearing in the moon, or were a town or city extending its boundaries, or were operations of this description carrying on, in a district where no such edifices had previously been erected, such objects and operations might probably be detected by a minute inspection. Were a multitude of living creatures moving from place to place, in a body, or were they even encamping in an extensive plain, like a large army, or like a tribe of Arabs in the desert, and afterwards removing, it is possible such changes might be traced by the difference of shade or color, which such movements would produce. In order to detect such minute objects and operations, it would be requisite that the surface of the moon should be distributed among at least a hundred astronomers, each having a spot or two allotted to him, as the object of his more particular investigation, and that the observations be continued for a period of at least thirty or forty years, during which time certain changes would probably be perceived, arising either from physical causes, or from the operations of living agents."[9]

LETTER XVI.

THE MOON.—PHASES.—HARVEST MOON.— LIBRATIONS.

"First to the neighboring Moon this mighty key
Of nature he applied. Behold! it turned
The secret wards, it opened wide the course
And various aspects of the queen of night:
Whether she wanes into a scanty orb,
Or, waxing broad, with her pale shadowy light,
In a soft deluge overflows the sky."—*Thomson's Elegy.*

LET us now inquire into the revolutions of the moon around the earth, and the various changes she undergoes every month, called her *phases*, which depend on the different positions she assumes, with respect to the earth and the sun, in the course of her revolution.

The moon revolves about the earth from west to east. Her apparent orbit, as traced out on the face of the sky, is a great circle; but this fact would not certainly prove that the orbit is really a circle, since, if it were an ellipse, or even a more irregular curve, the projection of it on the face of the sky would be a circle, as explained to you before. (See page 148.) The moon is comparatively so near to the earth, that her apparent movements are very rapid, so that, by attentively watching her progress in a clear night, we may see her move from star to star, changing her place perceptibly, every few hours. The interval during which she goes through the entire circuit of the heavens, from any star until she comes round to the same star again, is called a *sidereal month*, and consists of about twenty-seven and one fourth days. The time which intervenes between one new moon and another is called a *synodical month*, and consists of nearly twenty-nine and a half days. A new moon occurs when the sun and moon meet in the same part of the heavens; but the sun as well as the moon is apparently travelling eastward, and nearly at the rate of one degree a day, and consequently, during the twenty-seven days while the moon has been going round the earth, the sun has been going forward about the same number of degrees in the same direction. Hence, when the moon comes round to the part of the heavens where she passed the sun last, she does not find him there, but must go on more than two days, before she comes up with him again.

The moon does not pursue precisely the same track around the earth as the sun does, in his apparent annual motion, though she never deviates far from that track. The inclination of her orbit to the ecliptic is only about five degrees, and of course the moon is never seen further from the ecliptic than about that distance, and she is commonly much nearer to the ecliptic than five degrees. We may therefore see nearly what is the situation of the ecliptic in our evening sky at any particular time of year, just by watching the path which the moon pursues, from night to night, from new to full moon.

The two points where the moon's orbit crosses the ecliptic are called her *nodes*. They are the intersections of the lunar and solar orbits, as the equinoxes are the intersections of the equinoctial and ecliptic, and, like the latter, are one hundred and eighty degrees apart.

The changes of the moon, commonly called her *phases*, arise from different portions of her illuminated side being turned towards the earth at different times. When the moon is first seen after the setting sun, her form is that of a bright crescent, on the side of the disk next to the sun, while the other portions of the disk shine with a feeble light, reflected to the moon from the earth. Every night, we observe the moon to be further and further eastward of the sun, until, when she has reached an elongation from the sun of ninety degrees, half her visible disk is enlightened, and she is said to be in her *first quarter*. The terminator, or line which separates the illuminated from the dark part of the moon, is convex towards the sun from the new to the first quarter, and the moon is said to be *horned*. The extremities of the crescent are called *cusps*. At the first quarter, the terminator becomes a straight line, coinciding with the diameter of the disk; but after passing this point, the terminator becomes concave towards the sun, bounding that side of the moon by an elliptical curve, when the moon is said to be *gibbous*. When the moon arrives at the distance of one hundred and eighty degrees from the sun, the entire circle is illuminated, and the moon is *full*. She is then *in opposition* to the sun, rising about the time the sun sets. For a week after the full, the moon appears gibbous again, until, having arrived within ninety degrees of the sun, she resumes the same form as at the first quarter, being then at her *third quarter*. From this time until new moon, she exhibits again the form of a crescent before the rising sun, until, approaching her *conjunction* with the sun, her narrow thread of light is lost in the solar blaze; and finally, at the moment of passing the sun, the dark side is wholly turned towards us, and for some time we lose sight of the moon.

By inspecting Fig. 38, (where T represents the earth, A, B, C, &c., the moon in her orbit, and *a, b, c*, &c., her phases, as seen in the heavens,) we shall easily see how all these changes occur.

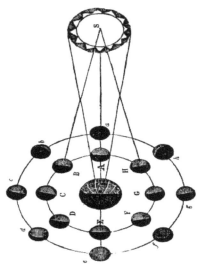

Fig. 38.

You have doubtless observed, that the moon appears much further in the south at one time than at another, when of the same age. This is owing to the fact that the ecliptic, and of course the moon's path, which is always very near it, is differently situated with respect to the *horizon*, at a given time of night, at different seasons of the year. This you will see at once, by turning to an artificial globe, and observing how the ecliptic stands with respect to the horizon, at different periods of the revolution. Thus, if we place the two equinoctial points in the eastern and western horizon, Libra being in the west, it will represent the position of the ecliptic at sunset in the month of September, when the sun is crossing the equator; and at that season of the year, the moon's path through our evening sky, one evening after another, from new to full, will be nearly along the same route, crossing the meridian nearly at right angles. But if we place the Winter solstice, or first degree of Capricorn, in the western horizon, and the first degree of Cancer in the eastern, then the position of the ecliptic will be very oblique to the meridian, the Winter solstice being very far in the southwest, and the Summer solstice very far in the northeast; and the course of the moon from new to full will be nearly along this track. Keeping these things in mind, we may easily see why the moon runs sometimes high and sometimes low. Recollect, also, that the new moon is always in the same part of the heavens with the sun, and that the full moon is in the opposite part of the heavens from the sun. Now, when the sun is at the Winter solstice, it sets far in the southwest, and accordingly the new moon runs very low; but the full moon, being in the opposite tropic, which rises far in the northeast, runs very high, as is known to be the case in mid-winter. But now take the position of the ecliptic in mid-summer. Then, at sunset, the tropic of Cancer is in the northwest, and the tropic of Capricorn in the southeast; consequently, the new moons run high and the full moons low.

It is a natural consequence of this arrangement, to render the moon's light the most beneficial to us, by giving it to us in greatest abundance, when we have least of the sun's light, and giving it to us most sparingly, when the sun's light is greatest. Thus, during the long nights of Winter, the full moon runs high, and continues a very long time above the horizon; while in mid-summer, the full moon runs low, and is above the horizon for a much shorter period. This arrangement operates very favorably to the inhabitants of the polar regions. At the season when the sun is absent, and they have constant night, then the moon, during the second and third quarters, embracing the season of full moon, is continually above the horizon, compensating in no small degree for the absence of the sun; while, during the Summer months, when the sun is constantly above the horizon, and the light of the moon is not needed, then she is above the horizon during the first and last quarters, when her light is least, affording at that time her greatest light to the inhabitants of the other hemisphere, from whom the sun is withdrawn.

About the time of the Autumnal equinox, the moon, when near her full, rises about sunset a number of nights in succession. This occasions a remarkable number of brilliant moonlight evenings; and as this is, in England, the period of harvest, the phenomenon is called the *harvest moon*. Its return is celebrated, particularly among the peasantry, by festive dances, and kept as a festival, called the *harvest home,*—an occasion often alluded to by the British poets. Thus Henry Kirke White:

"Moon of harvest, herald mild
Of plenty, rustic labor's child,
Hail, O hail! I greet thy beam,
As soft it trembles o'er the stream,
And gilds the straw-thatch'd hamlet wide,
Where innocence and peace reside;
Tis thou that glad'st with joy the rustic throng,
Promptest the tripping dance, th' exhilarating song."

To understand the reason of the harvest moon, we will, as before, consider the moon's orbit as coinciding with the ecliptic, because we may then take the ecliptic, as it is drawn on the artificial globe, to represent that orbit. We will also bear in mind, (what has been fully illustrated under the last head,) that, since the ecliptic cuts the meridian obliquely, while all the circles of diurnal revolution cut it perpendicularly, different portions of the ecliptic will cut the horizon at different angles. Thus, when the equinoxes are in the horizon, the ecliptic makes a very small angle with the horizon; whereas, when the solstitial points are in the horizon, the same angle is far greater. In the former case, a body moving eastward in the ecliptic, and being at the eastern horizon at sunset, would descend but a little way below the horizon in moving over many degrees of the ecliptic. Now, this is just the case of the moon at the time of the harvest home, about the time of the Autumnal equinox. The sun being then in Libra, and the moon, when full, being of course opposite to the sun, or in Aries; and moving eastward, in or near the ecliptic, at the rate of about thirteen degrees per day, would descend but a small distance below the horizon for five or six days in succession; that is for two or three days before, and the same number of days after, the full; and would consequently rise during all these evenings nearly at the same time, namely, a little before, or a little after, sunset, so as to afford a remarkable succession of fine moonlight evenings.

The moon *turns on her axis* in the same time in which she revolves around the earth. This is known by the moon's always keeping nearly the same face towards us, as is indicated by the telescope, which could not happen unless her revolution on her axis kept pace with her motion in her orbit. Take an apple, to represent the moon; stick a knittingneedle through it, in the direction of the stem, to represent the axis, in which case the two eyes of the apple will aptly represent the poles. Through the poles cut a line around the apple, dividing it into two hemispheres, and mark them, so as to be readily distinguished from each other. Now place a candle on the table, to represent the earth, and holding the apple by the knittingneedle, carry it round the candle, and you will see that, unless you make the apple turn round on the axis as you carry it about the candle, it will present different sides towards the candle; and that, in order to make it always present the same side, it will be necessary to make it revolve exactly once on its axis, while it is going round the circle,—the revolution on its axis always keeping exact pace with the motion in its orbit. The same thing will be observed, if you walk around a tree, always keeping your face towards the tree. If you have your face towards the tree when you set out, and walk round without turning, when you have reached the opposite side of the tree, your back will be towards it, and you will find that, in order to keep your face constantly towards the tree, it will be necessary to turn yourself round on your heel at the same rate as you go forward.

Since, however, the motion of the moon on its axis is uniform, while the motion in its orbit is unequal, the moon does in fact reveal to us a little sometimes of one side and sometimes of the other. Thus if, while carrying the apple round the candle, you carry it forward a little faster than the rate at which it turns on its axis, a portion of the hemisphere usually out of sight is brought into view on one side; or if the apple is moved forward slower than it is turned on its axis, a portion of the same hemisphere comes into view on the other side. These appearances are called the moon's *librations in longitude*. The moon has also a *libration in latitude*;—so called, because in one part of her revolution more of the region around one of the poles comes into view, and, in another part of the revolution, more of the region around the other pole, which gives the appearance of a tilting motion to the moon's axis. This is owing to the fact, that the moon's axis is inclined to the plane of her orbit. If, in the experiment with the apple, you hold the knittingneedle parallel to the candle, (in which case the axis will be perpendicular to the plane of revolution,) the candle will shine upon both poles during the whole circuit, and an eye situated where the candle is would constantly see both poles; but now incline the needle towards the plane of revolution, and carry it round, always keeping it parallel to itself, and you will observe that the two poles will be alternately in and out of sight.

The moon exhibits another appearance of this kind, called her *diurnal libration*, depending on the daily rotation of the spectator. She turns the same face towards the *centre* of the earth only, whereas we view her from the surface. When she is on the meridian, we view her disk nearly as though we viewed it from the centre of the earth, and hence, in this situation, it is subject to little change; but when she is near the horizon, our circle of vision takes in more of the upper limb than would be presented to a spectator at the centre of the earth. Hence, from this cause, we see a portion of one limb while the moon is rising, which is gradually lost sight of, and we see a portion of the opposite limb, as the moon declines to the west. You will remark that neither of the foregoing changes implies any actual motion in the moon, but that each arises from a change of position in the spectator. Since the succession of day and night depends on the revolution of a planet on its own axis, and it takes the moon twenty-nine and a half days to perform this revolution, so that the sun shall go from the meridian of any place and return to the same meridian again, of course the lunar day occupies this long period. So protracted an exposure to the sun's rays, especially in the equatorial regions of the moon, must occasion an excessive accumulation of heat; and so long an absence of the sun must occasion a corresponding degree of cold. A spectator on the side of the moon which is opposite to us would never see the earth, but one on the side next to us would see the earth constantly in his firmament, undergoing a gradual succession of changes, corresponding to those which the moon exhibits to the earth, but in the reverse order. Thus, when it is full moon to us, the earth, as seen from the moon, is then in conjunction with the sun, and of course presents her dark side to the moon.

Soon after this, an inhabitant of the moon would see a crescent, resembling our new moon, which would in like manner increase and go through all the changes, from new to full, and from full to new, as we see them in the moon. There are, however, in the two cases, several striking points of difference. In the first place, instead of twenty-nine and a half days, all these changes occur in one lunar day and night. During the first and last quarters, the changes would occur in the day-time; but during the second and third quarters, during the night. By this arrangement, the lunarians would enjoy the greatest possible benefit from the light afforded by the earth, since in the half of her revolution where she appears to them as full, she would be present while the sun was absent, and would afford her least light while the sun was present. In the second place, the earth would appear thirteen times as large to a spectator on the moon as the moon appears to us, and would afford nearly the same proportion of light, so that their long nights must be continually cheered by an extraordinary degree of light derived from this source; and if the full moon is hailed by our poets as "refulgent lamp of night,"[10] with how much more reason might a lunarian exult thus, in view of the splendid orb that adorns his nocturnal sky! In the third place, the earth, as viewed from any particular place on the moon, would occupy invariably the same part of the heavens. For while the rotation of the moon on her axis from west to east would appear to make the earth (as the moon does to us) revolve from east to west, the corresponding progress of the moon in her orbit would make the earth appear to revolve from west to east; and as these two motions are equal, their united effect would be to keep the moon apparently stationary in the sky. Thus, a spectator at E, Fig. 38, page 175, in the middle of the disk that is turned towards the earth, would have the earth constantly on his meridian, and at E, the conjunction of the earth and sun would occur at mid-day; but when the moon arrived at G, the same place would be on the margin of the circle of illumination, and will have the sun in the horizon; but the earth would still be on his meridian and in quadrature. In like manner, a place situated on the margin of the circle of illumination, when the moon is at E, would have the earth in the horizon; and the same place would always see the earth in the horizon, except the slight variations that would occur from the librations of the moon. In the fourth place, the earth would present to a spectator on the moon none of that uniformity of aspect which the moon presents to us, but would exhibit an appearance exceedingly diversified. The comparatively rapid rotation of the earth, repeated fifteen times during a lunar night, would present, in rapid succession, a view of our seas, oceans, continents, and mountains, all diversified by our clouds, storms, and volcanoes.

LETTER XVII.

MOON'S ORBIT.—HER IRREGULARITIES.

"Some say the zodiac constellations
Have long since left their antique stations,
Above a sign, and prove the same
In Taurus now, once in the Ram;
That in twelve hundred years and odd,
The sun has left his ancient road,
And nearer to the earth is come,'
Bove fifty thousand miles from home."—*Hudibras.*

WE have thus far contemplated the revolution of the moon around the earth as though the earth were at rest. But in order to have just ideas respecting the moon's motions, we must recollect that the moon likewise revolves along with the earth around the sun. It is sometimes said that the earth *carries* the moon along with her, in her annual revolution. This language may convey an erroneous idea; for the moon, as well as the earth, revolves around the sun under the influence of two forces, which are independent of the earth, and would continue her motion around the sun, were the earth removed out of the way. Indeed, the moon is attracted towards the sun two and one fifth times more than towards the earth, and would abandon the earth, were not the latter also carried along with her by the same forces. So far as the sun acts equally on both bodies, the motion with respect to each other would not be disturbed. Because the gravity of the moon towards the sun is found to be greater, at the conjunction, than her gravity towards the earth, some have apprehended that, if the doctrine of universal gravitation is true, the moon ought necessarily to abandon the earth. In order to understand the reason why it does not do thus, we must reflect, that, when a body is revolving in its orbit under the influence of the projectile force and gravity, whatever diminishes the force of gravity, while that of projection remains the same, causes the body to approach nearer to the tangent of her orbit, and of course to recede from the centre; and whatever increases the amount of gravity, carries the body towards the centre. Thus, in Fig. 33, 152, if, with a certain force of projection acting in the direction A B, and of attraction, in the direction A C, the attraction which caused a body to move in the line A D were diminished, it would move nearer to the tangent, as in A E, or A F. Now, when the moon is in conjunction, her gravity towards the earth acts in opposition to that towards the sun, (see Fig. 38, page 175,) while her velocity remains too great to carry her with what force remains, in a circle about the sun, and she therefore recedes from the sun, and commences her revolution around the earth. On arriving at the opposition, the gravity of the earth conspires with that of the sun, and the moon's projectile force being less than that required to make her revolve in a circular orbit, when attracted towards the sun by the sum of these forces, she accordingly begins to approach the sun, and descends again to the conjunction.

The attraction of the sun, however, being every where greater than that of the earth, the actual path of the moon around the sun is every where concave towards the latter. Still, the elliptical path of the moon around the earth is to be conceived of, in the same way as though both bodies were at rest with respect to the sun. Thus, while a steam-boat is passing *swiftly* around an island, and a man is walking *slowly* around a post in the cabin, the line which he describes in space between the forward motion of the boat and his circular motion around the post, may be every where concave towards the island, while his path around the post will still be the same as though both were at rest. A nail in the rim of a coach-wheel will turn around the axis of the wheel, when the coach has a forward motion, in the same manner as when the coach is at rest, although the line actually described by the nail will be the resultant of both motions, and very different from either.

We have hitherto regarded the moon as describing a great circle on the face of the sky, such being the visible orbit, as seen by projection. But, on a more exact investigation, it is found that her orbit is not a circle, and that her motions are subject to very numerous irregularities. These will be best understood in connexion with the causes on which they depend. The law of universal gravitation has been applied with wonderful success to their developement, and its results have conspired with those of long-continued observation, to furnish the means of ascertaining with great exactness the place of the moon in the heavens, at any given instant of time, past or future, and thus to enable astronomers to determine longitudes, to calculate eclipses, and to solve other problems of the highest interest. The whole number of irregularities to which the moon is subject is not less than sixty, but the greater part are so small as to be hardly deserving of attention; but as many as thirty require to be estimated and allowed for, before we can ascertain the exact place of the moon at any given time. You will be able to understand something of the cause of these irregularities, if you first gain a distinct idea of the mutual actions of the sun, the moon, and the earth. The irregularities in the moon's motions are due chiefly to the disturbing influence of the sun, which operates in two ways; first, by acting unequally on the earth and moon; and secondly, by acting obliquely on the moon, on account of the inclination of her orbit to the ecliptic. If the sun acted equally on the earth and moon, and always in parallel lines, this action would serve only to restrain them in their annual motions around the sun, and would not affect their actions on each other, or their motions about their common centre of gravity. In that case, if they were allowed to fall towards the sun, they would fall equally, and their respective situations would not be affected by their descending equally towards it. But, because the moon is nearer the sun in one half of her orbit than the earth is, and in the other half of her orbit is at a greater distance than the earth from the sun, while the power of gravity is always greater at a less distance; it follows, that in one half of her orbit the moon is more attracted than the earth towards the sun, and, in the other half, less attracted than the earth.

To see the effects of this process, let us suppose that the projectile motions of the earth and moon were destroyed, and that they were allowed to fall freely towards the sun. (See Fig. 38, page 175.) If the moon was in conjunction with the sun, or in that part of her orbit which is nearest to him, the moon would be more attracted than the earth, and fall with greater velocity towards the sun; so that the distance of the moon from the earth would be increased by the fall. If the moon was in opposition, or in the part of her orbit which is furthest from the sun, she would be less attracted than the earth by the sun, and would fall with a less velocity, and be left behind; so that the distance of the moon from the earth would be increased in this case, also. If the moon was in one of the quarters, then the earth and the moon being both attracted towards the centre of the sun, they would both descend directly towards that centre, and, by approaching it, they would necessarily at the same time approach each other, and in this case their distance from each other would be diminished. Now, whenever the action of the sun would increase their distance, if they were allowed to fall towards the sun, then the sun's action, by endeavoring to separate them, diminishes their gravity to each other; whenever the sun's action would diminish the distance, then it increases their mutual gravitation. Hence, in the conjunction and opposition, their gravity towards each other is diminished by the action of the sun, while in the quadratures it is increased. But it must be remembered, that it is not the total action of the sun on them that disturbs their motions, but only that part of it which tends at one time to separate them, and at another time to bring them nearer together. The other and far greater part has no other effect than to retain them in their annual course around the sun.

The cause of the lunar irregularities was first investigated by Sir Isaac Newton, in conformity with his doctrine of universal gravitation, and the explanation was first published in the 'Principia;' but, as it was given in a mathematical dress, there were at that age very few persons capable of reading or understanding it. Several eminent individuals, therefore, undertook to give a popular explanation of these difficult points. Among

Newton's contemporaries, the best commentator was M'Laurin, a Scottish astronomer, who published a large work entitled 'M'Laurin's Account of Sir Isaac Newton's Discoveries.' No writer of his own day, and, in my opinion, no later commentator, has equalled M'Laurin, in reducing to common apprehension the leading principles of the doctrine of gravitation, and the explanation it affords of the motions of the heavenly bodies. To this writer I am indebted for the preceding easy explanation of the irregularities of the moon's motions, as well as for several other illustrations of the same sublime doctrine.

The figure of the moon's orbit is an ellipse. We have before seen, that the earth's orbit around the sun is of the same figure; and we shall hereafter see this to be true of all the planetary orbits. The path of the earth, however, departs very little from a circle; that of the moon differs materially from a circle, being considerably longer one way than the other. Were the orbit a circle having the earth in the centre, then the radius vector, or line drawn from the centre of the moon to the centre of the earth, would always be of the same length; but it is found that the length of the radius vector is only fifty-six times the radius of the earth when the moon is nearest to us, while it is sixty-four times that radius when the moon is furthest from us. The point in the moon's orbit nearest the earth is called her *perigee*; the point furthest from the earth, her *apogee*. We always know when the moon is at one of these points, by her apparent diameter or apparent velocity; for, when at the perigee, her diameter is greater than at any time, and her motion most rapid; and, on the other hand, her diameter is least, and her motion slowest, when she is at her apogee.

The moon's nodes constantly shift their positions in the ecliptic, from east to west, at the rate of about nineteen and a half degrees every year, returning to the same points once in eighteen and a half years. In order to understand what is meant by this backward motion of the nodes, you must have very distinctly in mind the meaning of the terms themselves; and if, at any time, you should be at a loss about the signification of any word that is used in expressing an astronomical proposition, I would advise you to turn back to the previous definition of that term, and revive its meaning clearly in the mind, before you proceed any further. In the present case, you will recollect that the moon's nodes are the two points where her orbit cuts the plane of the ecliptic. Suppose the great circle of the ecliptic marked out on the face of the sky in a distinct line, and let us observe, at any given time, the exact moment when the moon crosses this line, which we will suppose to be close to a certain star; then, on its next return to that part of the heavens, we shall find that it crosses the ecliptic sensibly to the westward of that star, and so on, further and further to the westward, every time it crosses the ecliptic at either node. This fact is expressed by saying that *the nodes retrograde on the ecliptic*; since any motion from east to west, being contrary to the order of the signs, is called retrograde. The line which joins these two points, or the line of the nodes, is also said to have a retrograde motion, or to revolve from east to west once in eighteen and a half years.

The *line of the apsides* of the moon's orbit revolves from west to east, through her whole course, in about nine years. You will recollect that the apsides of an elliptical orbit are the two extremities of the longer axis of the ellipse; corresponding to the perihelion and aphelion of bodies revolving about the sun, or to the perigee and apogee of a body revolving about the earth. If, in any revolution of the moon, we should accurately mark the place in the heavens where the moon is nearest the earth, (which may be known by the moon's apparent diameter being then greatest,) we should find that, at the next revolution, it would come to its perigee a little further eastward than before, and so on, at every revolution, until, after nine years, it would come to its perigee nearly at the same point as at first. This fact is expressed by saying, that the perigee, and of course the apogee, revolves, and that the line which joins these two points, or the line of the apsides, also revolves.

These are only a few of the irregularities that attend the motions of the moon. These and a few others were first discovered by actual observation and have been long known; but a far greater number of lunar irregularities have been made known by following out all the consequences of the law of universal gravitation.

The moon may be regarded as a body endeavoring to make its way around the earth, but as subject to be continually impeded, or diverted from its main course, by the action of the sun and of the earth; sometimes acting in concert and sometimes in opposition to each other. Now, by exactly estimating the amount of these respective forces, and ascertaining their resultant or combined effect, in any given case, the direction and velocity of the moon's motion may be accurately determined. But to do this has required the highest powers of the human mind, aided by all the wonderful resources of mathematics. Yet, so consistent is truth with itself, that, where some minute inequality in the moon's motions is developed at the end of a long and intricate mathematical process, it invariably happens, that, on pointing the telescope to the moon, and watching its progress through the skies, we may actually see her commit the same irregularities, unless (as is the case with many of them) they are too minute to be matters of observation, being beyond the powers of our vision, even when aided by the best telescopes. But the truth of the law of gravitation, and of the results it gives, when followed out by a chain of mathematical reasoning, is fully confirmed, even in these minutest matters, by the fact that the moon's place in the heavens, when thus determined, always corresponds, with wonderful exactness, to the place which she is actually observed to occupy at that time.

The mind, that was first able to elicit from the operations of Nature the law of universal gravitation, and afterwards to apply it to the complete explanation of all the irregular wanderings of the moon, must have given evidence of intellectual powers far elevated above those of the majority of the human race. We need not wonder, therefore, that such homage is now paid to the genius of Newton,—an admiration which has been continually increasing, as new discoveries have been made by tracing out new consequences of the law of universal gravitation.

The chief object of astronomical *tables* is to give the amount of all the irregularities that attend the motions of the heavenly bodies, by estimating the separate value of each, under all the different circumstances in which a body can be placed. Thus, with respect to the moon, before we can determine accurately the distance of the moon from the vernal equinox, that is, her longitude at any given moment, we must be able to make exact allowances for all her irregularities which would affect her longitude. These are in all no less than sixty, though most of them are so exceedingly minute, that it is not common to take into the account more than twenty-eight or thirty. The values of these are all given in the lunar tables; and in finding the moon's place, at any given time, we proceed as follows: We first find what her place would be on the supposition that she moves uniformly in a circle. This gives her *mean* place. We next apply the various corrections for her irregular motions; that is, we apply the *equations*, subtracting some and adding others, and thus we find her *true* place.

The astronomical tables have been carried to such an astonishing degree of accuracy, that it is said, by the highest authority, that an astronomer could now predict, for a thousand years to come, the precise moment of the passage of any one of the stars over the meridian wire of the telescope of his transit-instrument, with such a degree of accuracy, that the error would not be so great as to remove the object through an angular space corresponding to the semidiameter of the finest wire that could be made; and a body which, by the tables, ought to appear in the transit-instrument in the middle of that wire, would in no case be removed to its outer edge. The astronomer, the mathematician, and the artist, have united their powers to produce this great result. The astronomer has collected the data, by long-continued and most accurate observations on the actual motions of the heavenly bodies, from night to night, and from year to year; the mathematician has taken these data, and applied to them the boundless resources of geometry and the calculus; and, finally, the instrument-maker has furnished the means, not only of verifying these conclusions, but of discovering new truths, as the foundation of future reasonings.

Since the points where the moon crosses the ecliptic, or the moon's nodes, constantly shift their positions about nineteen and a half degrees to the westward, every year, the sun, in his annual progress in the ecliptic, will go from the node round to the same node again in less time than a year, since the node goes to meet him nineteen and a half degrees to the west of the point where they met before. It would have taken the sun about nineteen days to have passed over this arc; and consequently, the interval between two successive conjunctions between the sun and the moon's node is about nineteen days shorter than the solar year of three hundred and sixty-five days; that is, it is about three hundred and forty-six days; or, more exactly, it is 346.619851 days. The time from one new moon to another is 29.5305887 days. Now, nineteen of the former periods are almost exactly equal to two hundred and twenty-three of the latter:

For $346.619851 \times 19 = 6585.78$ days $= 18$ y. 10 d.

And $29.5305887 \times 223 = 6585.32$ " $=$ " " " "

Hence, if the sun and moon were to leave the moon's node together, after the sun had been round to the same node nineteen times, the moon would have made very nearly two hundred and twenty-three conjunctions with the sun. If, therefore, she was in conjunction with the sun at the beginning of this period, she would be in conjunction again at the end of it; and all things relating to the sun, the moon, and the node, would be restored to the same relative situation as before, and the sun and moon would start again, to repeat the same phenomena, arising out of these relations, as occurred in the preceding period, and in the same order. Now, when the sun and moon meet at the moon's node, an eclipse of the sun happens; and during the entire period of eighteen and a half years eclipses will happen, nearly in the same manner as they did at corresponding times in the preceding period. Thus, if there was a great eclipse of the sun on the fifth year of one of these periods, a similar eclipse (usually differing somewhat in magnitude) might be expected on the fifth year of the next period. Hence this period, consisting of about eighteen years and ten days, under the name of the *Saros*, was used by the Chaldeans, and other ancient nations, in predicting eclipses. It was probably by this means that Thales, a Grecian astronomer who flourished six hundred years before the Christian era, predicted an eclipse of the sun. Herodotus, the old historian of Greece, relates that the day was suddenly changed into night, and that Thales of Miletus had foretold that a great eclipse was to happen *this year*. It was therefore, at that age, considered as a distinguished feat to predict even the year in which an eclipse was to happen. This eclipse is memorable in ancient history, from its having terminated the war between the Lydians and the Medes, both parties being smitten with such indications of the wrath of the gods.

The *Metonic Cycle* has sometimes been confounded with the Saros, but it is not the same with it, nor was the period used, like the Saros, for foretelling eclipses, but for ascertaining the *age* of the moon at any given period. It consisted of nineteen tropical years, during which time there are exactly two hundred and thirty-five new moons; so that, at the end of this period, the new moons will recur at seasons of the year corresponding exactly to those of the preceding cycle. If, for example, a new moon fell at the time of the vernal equinox, in one cycle, nineteen years afterwards it would occur again at the same equinox; or, if it had happened ten days after the equinox, in one cycle, it would also happen ten days after the equinox, nineteen years afterwards. By registering, therefore, the exact days of any cycle at which the new or full moons occurred, such a calendar would show on what days these events would occur in any other cycle; and, since the regulation of games, feasts, and fasts, has been made very extensively, both in ancient and modern times, according to new or full moons, such a calendar becomes very convenient for finding the day on which the new or full moon required takes place. Suppose, for example, it were decreed that a festival should be held on the day of the first full moon after the Vernal equinox. Then, to find on what day that would happen, in any given year, we have only to see what year it is of the lunar cycle; for the day will be the same as it was in the corresponding year of the calendar which records all the full moons of the cycle for each year, and the respective days on which they happen.

The Athenians adopted the metonic cycle four hundred and thirty-three years before the Christian era, for the regulation of their calendars, and had it inscribed in letters of gold on the walls of the temple of Minerva. Hence the term *golden number*, still found in our almanacs, which denotes the year of the lunar cycle. Thus, fourteen was the golden number for 1837, being the fourteenth year of the lunar cycle.

The inequalities of the moon's motions are divided into periodical and secular. *Periodical* inequalities are those which are completed in comparatively short periods. *Secular* inequalities are those which are completed only in very long periods, such as centuries or ages. Hence the corresponding terms *periodical equations* and *secular equations*. As an example of a secular inequality, we may mention the acceleration of the *moon's mean motion*. It is discovered that the moon actually revolves around the earth in a less period now than she did in ancient times. The difference, however, is exceedingly small, being only about ten seconds in a century. In a lunar eclipse, the moon's longitude differs from that of the sun, at the middle of the eclipse, by exactly one hundred and eighty degrees; and since the sun's longitude at any given time of the year is known, if we can learn the day and hour when an eclipse occurred at any period of the world, we of course know the longitude of the sun and moon at that period. Now, in the year 721, before the Christian era, Ptolemy records a lunar eclipse to have happened, and to have been observed by the Chaldeans. The moon's longitude, therefore, for that time, is known; and as we know the mean motions of the moon, at present, starting from that epoch, and computing, as may easily be done, the place which the moon ought to occupy at present, at any given time, she is found to be actually nearly a degree and a half in advance of that place. Moreover, the same conclusion is derived from a comparison of the Chaldean observations with those made by an Arabian astronomer of the tenth century.

This phenomenon at first led astronomers to apprehend that the moon encountered a resisting medium, which, by destroying at every revolution a small portion of her projectile force, would have the effect to bring her nearer and nearer to the earth, and thus to augment her velocity. But, in 1786, La Place demonstrated that this acceleration is one of the legitimate effects of the sun's disturbing force, and is so connected with changes in the eccentricity of the earth's orbit, that the moon will continue to be accelerated while that eccentricity diminishes; but when the eccentricity has reached its minimum, or lowest point, (as it will do, after many ages,) and begins to increase, then the moon's motions will begin to be retarded, and thus her mean motions will oscillate for ever about a mean value.

LETTER XVIII.

ECLIPSES.

——"As when the sun, new risen,
Looks through the horizontal misty air,
Shorn of his beams, or from behind the moon,
In dim eclipse, disastrous twilight sheds
On half the nations, and with fear of change
Perplexes monarchs: darkened so, yet shone,
Above them all, the Archangel."—*Milton*.

HAVING now learned various particulars respecting the earth, the sun, and the moon, you are prepared to understand the explanation of solar and lunar eclipses, which have in all ages excited a high degree of interest. Indeed, what is more admirable, than that astronomers should be able to tell us, years beforehand, the exact instant of the commencement and termination of an eclipse, and describe all the attendant circumstances with the greatest fidelity. You have doubtless, my dear friend, participated in this admiration, and felt a strong desire to learn how it is that astronomers are able to look so far into futurity. I will endeavor, in this Letter, to explain to you the leading principles of the calculation of eclipses, with as much plainness as possible.

An *eclipse of the moon* happens when the moon, in its revolution around the earth, falls into the earth's shadow. An *eclipse of the sun* happens when the moon, coming between the earth and the sun, covers either a part or the whole of the solar disk.

The earth and the moon being both opaque, globular bodies, exposed to the sun's light, they cast shadows opposite to the sun, like any other bodies on which the sun shines. Were the sun of the same size with the earth and the moon, then the lines drawn touching the surface of the sun and the surface of the earth or moon (which lines form the boundaries of the shadow) would be parallel to each other, and the shadow would be a cylinder infinite in length; and were the sun less than the earth or the moon, the shadow would be an increasing cone, its narrower end resting on the earth; but as the sun is vastly greater than either of these bodies, the shadow of each is a cone whose base rests on the body itself, and which comes to a point, or vertex, at a certain distance behind the body. These several cases are represented in the following diagrams, Figs. 39, 40, 41.

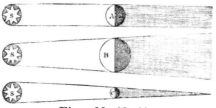

Figs. 39, 40, 41.

It is found, by calculation, that the length of the moon's shadow, on an average, is just about sufficient to reach to the earth; but the moon is sometimes further from the earth than at others, and when she is nearer than usual, the shadow reaches considerably beyond the surface of the earth. Also, the moon, as well as the earth, is at different distances from the sun at different times, and its shadow is longest when it is furthest from the sun. Now, when both these circumstances conspire, that is, when the moon is in her perigee and along with the earth in her aphelion, her shadow extends nearly fifteen thousand miles beyond the centre of the earth, and covers a space on the surface one hundred and seventy miles broad. The earth's shadow is nearly a million of miles in length, and consequently more than three and a half times as long as the distance of the earth from the moon; and it is also, at the distance of the moon, three times as broad as the moon itself.

An eclipse of the sun can take place only at new moon, when the sun and moon meet in the same part of the heavens, for then only can the moon come between us and the sun; and an eclipse of the moon can occur only when the sun and moon are in opposite parts of the heavens, or at full moon; for then only can the moon fall into the shadow of the earth.

Fig. 42.

The nature of eclipses will be clearly understood from the following representation. The diagram, Fig. 42, exhibits the relative position of the sun, the earth, and the moon, both in a solar and in a lunar eclipse. Here, the moon is first represented, while revolving round the earth, as passing between the earth and the sun, and casting its shadow on the earth. As the moon is here supposed to be at her average distance from the earth, the shadow but just reaches the earth's surface. Were the moon (as is sometimes the case) nearer the earth her shadow would not terminate in a point, as is represented in the figure, but at a greater or less distance nearer the base of the cone, so as to cover a considerable space, which, as I have already mentioned, sometimes extends to one hundred and seventy miles in breadth, but is commonly much less than this. On the other side of the earth, the moon is represented as traversing the earth's shadow, as is the case in a lunar eclipse. As the moon is sometimes nearer the earth and sometimes further off, it is evident that it will traverse the shadow at a broader or a narrower part, accordingly. The figure, however, represents the moon as passing the shadow further from the earth than is ever actually the case, since the distance from the earth is never so much as one third of the whole length of the shadow.

It is evident from the figure, that if a spectator were situated where the moon's shadow strikes the earth, the moon would cut off from him the view of the sun, or the sun would be totally eclipsed. Or, if he were within a certain distance of the shadow on either side, the moon would be partly between him and the sun, and would intercept from him more or less of the sun's light, according as he was nearer to the shadow or further from it. If he were at *c* or *d*, he would just see the moon entering upon the sun's disk; if he were nearer the shadow than either of these points, he would have a portion of this light cut off from his view, and more, in proportion as he drew nearer the shadow; and the moment he entered the shadow, he would lose sight of the sun. To all places between *a* or *b* and the shadow, the sun would cast a partial shadow of the moon, growing deeper and deeper, as it approached the true shadow. This partial shadow is called the moon's *penumbra*. In like manner, as the moon approaches the earth's shadow, in a lunar eclipse, as soon as she arrives at *a*, the earth begins to intercept from her a portion of the sun's light, or she falls in the earth's penumbra. She continues to lose more and more of the sun's light, as she draws near to the shadow, and hence her disk becomes gradually obscured, until it enters the shadow, when the sun's light is entirely lost.

As the sun and earth are both situated in the plane of the ecliptic, if the moon also revolved around the earth in this plane, we should have a solar eclipse at every new moon, and a lunar eclipse at every full moon; for, in the former case, the moon would come directly between us and the sun, and in the latter case, the earth would come directly between the sun and the moon. But the moon is inclined to the ecliptic about five degrees, and the centre of the moon may be all this distance from the centre of the sun at new moon, and the same distance from the centre of the earth's shadow at full moon. It is true, the moon extends across her path, one half her breadth lying on each side of it, and the sun likewise reaches from the ecliptic a distance equal to half his breadth. But these luminaries together make but little more than a degree, and consequently, their two semidiameters would occupy only about half a degree of the five degrees from one orbit to the other where they are furthest apart. Also, the earth's shadow, where the moon crosses it, extends from the ecliptic less than three fourths of a degree, so that the semidiameter of the moon and of the earth's shadow would together reach but little way across the space that may, in certain cases, separate the two luminaries from each other when they are in opposition. Thus, suppose we could take hold of the circle in the figure that represents the moon's orbit, (Fig. 42, page 197,) and lift the moon up five degrees above the plane of the paper, it is evident that the moon, as seen from the earth, would appear in the heavens five degrees above the sun, and of course would cut off none of his light; and it is also plain that the moon, at the full, would pass the shadow of the earth five degrees below it, and would suffer no eclipse. But in the course of the sun's apparent revolution round the earth once a year he is successively in every part of the ecliptic; consequently, the conjunctions and oppositions of the sun and moon may occur at any part of the ecliptic, and of course at the two points where the moon's orbit crosses the ecliptic,—that is, at the nodes; for the sun must necessarily come to each of these nodes once a year. If, then, the moon overtakes the sun just as she is crossing his path, she will hide more or less of his disk from us. Since, also, the earth's shadow is always directly opposite to the sun, if the sun is at one of the nodes, the shadow must extend in the direction of the other node, so as to lie directly across the moon's path; and if the moon overtakes it there, she will pass through it, and be eclipsed. Thus, in Fig. 43, let BN represent the sun's path, and AN, the moon's,—N being the place of the node; then it is evident, that if the two luminaries at new moon be so far from the node, that the distances between their centres is greater than their semidiameters, no eclipse can happen; but if that distance is less than this sum, as at E, F, then an eclipse will take place; but if the position be as at C, D, the two bodies will just touch one another. If A denotes the earth's shadow, instead of the sun, the same illustration will apply to an eclipse of the moon.

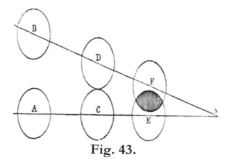

Fig. 43.

Since bodies are defined to be in conjunction when they are in the *same* part of the heavens, and to be in opposition when they are in *opposite* parts of the heavens, it may not appear how the sun and moon can be in conjunction, as at A and B, when they are still at some distance from each other. But it must be recollected that bodies are in conjunction when they have the same longitude, in which case they are situated in the same great circle perpendicular to the ecliptic,—that is, in the same secondary to the ecliptic. One of these bodies may be much further from the ecliptic than the other; still, if the same secondary to the ecliptic passes through them both, they will be in conjunction or opposition.

In a total eclipse of the moon, its disk is still visible, shining with a dull, red light. This light cannot be derived directly from the sun, since the view of the sun is completely hidden from the moon; nor by reflection from the earth, since the illuminated side of the earth is wholly turned from the moon; but it is owing to refraction from the earth's atmosphere, by which a few scattered rays of the sun are bent round into the earth's shadow and conveyed to the moon, sufficient in number to afford the feeble light in question.

It is impossible fully to understand the *method of calculating eclipses*, without a knowledge of trigonometry; still it is not difficult to form some general notion of the process. It may be readily conceived that, by long-continued observations on the sun and moon, the laws of their revolution may be so well understood, that the exact places which they will occupy in the heavens at any future times may be foreseen and laid down in tables of the sun and moon's motions; that we may thus ascertain, by inspecting the tables, the instant when these two bodies will be together in the heavens, or be in conjunction, and when they will be one hundred and eighty degrees apart, or in opposition. Moreover, since the exact place of the moon's node among the stars at any particular time is known to astronomers, it cannot be difficult to determine when the new or full moon occurs in the same part of the heavens as that where the node is projected, as seen from the earth. In short, as astronomers can easily determine what will be the relative position of the sun, the moon, and the moon's nodes, for any given time, they can tell when these luminaries will meet so near the node as to produce an eclipse of the sun, or when they will be in opposition so near the node as to produce an eclipse of the moon.

A little reflection will enable you to form a clear idea of the situation of the sun, the moon, and the earth, at the time of a solar eclipse. First, suppose the conjunction to take place at the node; that is, imagine the moon to come *directly* between the earth and the sun, as she will of course do, if she comes between the earth and the sun the moment she is crossing the ecliptic; for then the three bodies will all lie in one and the same straight line. But when the moon is in the ecliptic, her shadow, or at least the axis, or central line, of the shadow, must coincide with the line that joins the centres of the sun and earth, and reach along the plane of the ecliptic towards the earth. The moon's shadow, at her average distance from the earth, is just about long enough to reach the surface of the earth; but when the moon, at the new, is in her apogee, or at her greatest distance from the earth, the shadow is not long enough to reach the earth. On the contrary, when the moon is nearer to us than her average distance, her shadow is long enough to reach beyond the earth, extending, when the moon is in her perigee, more than fourteen thousand miles beyond the centre of the earth. Now, as during the eclipse the moon moves nearly in the plane of the ecliptic, her shadow which accompanies her must also move nearly in the same plane, and must therefore traverse the earth across its central regions, along the terrestrial ecliptic, since this is nothing more than the intersection of the plane of the celestial ecliptic with the earth's surface. The motion of the earth, too, on its axis, in the same direction, will carry a place along with the shadow, though with a less velocity by more than one half; so that the actual velocity of the shadow, in respect to places over which it passes on the earth, will only equal the difference between its own rate and that of the places, as they are carried forward in the diurnal revolution.

We have thus far supposed that the moon comes to her conjunction precisely at the node, or at the moment when she is crossing the ecliptic. But, secondly, suppose she is on the north side of the ecliptic at the time of conjunction, and moving towards her descending node, and that the conjunction takes place as far from the node as an eclipse can happen. The shadow will not fall in the plane of the ecliptic, but a little northward of it, so as just to graze the earth near the pole of the ecliptic. The nearer the conjunction comes to the node, the further the shadow will fall from the polar towards the equatorial regions.

In a solar eclipse, the shadow of the moon travels over a portion of the earth, as the shadow of a small cloud, seen from an eminence in a clear day, rides along over hills and plains. Let us imagine ourselves standing on the moon; then we shall see the earth partially eclipsed by the moon's shadow, in the same manner as we now see the moon eclipsed by the shadow of the earth; and we might calculate the various circumstances of the eclipse,—its commencement, duration, and quantity,—in the same manner as we calculate these elements in an eclipse of the moon, as seen from the earth. But although the general characters of a solar eclipse might be investigated on these principles, so far as respects the earth at large, yet, as the appearances of the same eclipse of the sun are very different at different places on the earth's surface, it is necessary to calculate its peculiar aspects for each place separately, a circumstance which makes the calculation of a solar eclipse much more complicated and tedious than that of an eclipse of the moon. The moon, when she enters the shadow of the earth, is deprived of the light of the part immersed, and the effect upon its appearance is the same as though that part were painted black, in which case it would be black alike to all places where the moon was above the horizon. But it not so with a solar eclipse. We do not see this by the shadow cast on the earth, as we should do, if we stood on the moon, but by the interposition of the moon between us and the sun; and the sun may be hidden from one observer, while he is in full view of another only a few miles distant. Thus, a small insulated cloud sailing in a clear sky will, for a few moments, hide the sun from us, and from a certain space near us, while all the region around is illuminated. But although the analogy between the motions of the shadow of a small cloud and of the moon in a solar eclipse holds good in many particulars, yet the velocity of the lunar shadow is far greater than that of the cloud, being no less than two thousand two hundred and eighty miles per hour.

The moon's shadow can never cover a space on the earth more than one hundred and seventy miles broad, and the space actually covered commonly falls much short of that. The portion of the earth's surface ever covered by the moon's penumbra is about four thousand three hundred and ninety-three miles.

The apparent diameter of the moon varies materially at different times, being greatest when the moon is nearest to us, and least when she is furthest off; while the sun's apparent dimensions remain nearly the same. When the moon is at her average distance from the earth, she is just about large enough to cover the sun's disk; consequently, if, in a central eclipse of the sun, the moon is at her mean distance, she covers the sun but for an instant, producing only a momentary eclipse. If she is nearer than her average distance, then the eclipse may continue total some time, though never more than eight minutes, and seldom so long as that; but if she is further off than usual, or towards her apogee, then she is not large enough to cover the whole solar disk, but we see a ring of the sun encircling the moon, constituting an *annular eclipse*, as seen in Fig. 44. Even the elevation of the moon above the horizon will sometimes sensibly affect the dimensions of the eclipse. You will recollect that the moon is nearer to us when on the meridian than when in the horizon by nearly four thousand miles, or by nearly the radius of the earth; and consequently, her apparent diameter is largest when on the meridian. The difference is so considerable, that the same eclipse will appear total to a spectator who views it near his meridian, while, at the same moment, it appears annular to one who has the moon near his horizon. An annular eclipse may last, at most, twelve minutes and twenty-four seconds.

Fig. 44.

Eclipses of the sun are more frequent than those of the moon. Yet lunar eclipses being visible to every part of the terrestrial hemisphere opposite to the sun, while those of the sun are visible only to a small portion of the hemisphere on which the moon's shadow falls, it happens that, for any particular place on the earth, lunar eclipses are more frequently visible than solar. In any year, the number of eclipses of both luminaries cannot be less than two nor more than seven: the most usual number is four, and it is very rare to have more than six. A total eclipse of the moon frequently happens at the next full moon after an eclipse of the sun. For since, in a solar eclipse, the sun is at or near one of the moon's nodes,—that is, is projected to the place in the sky where the moon crosses the ecliptic,—the earth's shadow, which is of course directly opposite to the sun, must be at or near the other node, and may not have passed too far from the node before the moon comes round to the opposition and overtakes it. In total eclipses of the sun, there has sometimes been observed a remarkable radiation of light from the margin of the sun, which is thought to be owing to the zodiacal light, which is of such dimensions as to extend far beyond the solar orb. A striking appearance of this kind was exhibited in the total eclipse of the sun which occurred in June, 1806.

A total eclipse of the sun is one of the most sublime and impressive phenomena of Nature. Among barbarous tribes it is ever contemplated with fear and astonishment, and as strongly indicative of the displeasure of the gods. Two ancient nations, the Lydians and Medes, alluded to before, who were engaged in a bloody war, about six hundred years before Christ, were smitten with such awe, on the appearance of a total eclipse of the sun, just on the eve of a battle, that they threw down their arms, and made peace. When Columbus first discovered America, and was in danger of hostility from the Natives, he awed them into submission by telling them that the sun would be darkened on a certain day, in token of the anger of the gods at them, for their treatment of him.

Among cultivated nations, a total eclipse of the sun is recognised, from the exactness with which the time of occurrence and the various appearances answer to the prediction, as affording one of the proudest triumphs of astronomy. By astronomers themselves, it is of course viewed with the highest interest, not only as verifying their calculations, but as contributing to establish, beyond all doubt, the certainty of those grand laws, the truth of which is involved in the result. I had the good fortune to witness the total eclipse of the sun of June, 1806, which was one of the most remarkable on record. To the wondering gaze of childhood it presented a spectacle that can never be forgotten. A bright and beautiful morning inspired universal joy, for the sky was entirely cloudless. Every one was busily occupied in preparing smoked glass, in readiness for the great sight, which was to be first seen about ten o'clock. A thrill of mingled wonder and delight struck every mind when, at the appointed moment, a little black indentation appeared on the limb of the sun. This gradually expanded, covering more and more of the solar disk, until an increasing gloom was spread over the face of Nature; and when the sun was wholly lost, near mid-day, a feeling of horror pervaded almost every beholder. The darkness was wholly unlike that of twilight or night. A thick curtain, very different from clouds, hung upon the face of the sky, producing a strange and indescribably gloomy appearance, which was reflected from all things on the earth, in hues equally strange and unnatural. Some of the planets, and the largest of the fixed stars, shone out through the gloom, yet with their usual brightness. The temperature of the air rapidly declined, and so sudden a chill came over the earth, that many persons caught severe colds from their exposure. Even the animal tribes exhibited tokens of fear and agitation. Birds, especially, fluttered and flew swiftly about, and domestic fowls went to rest.

Indeed, the word *eclipse* is derived from a Greek word, (εκλειψις, *ekleipsis,*) which signifies to fail, to faint or swoon away; since the moon, at the period of her greatest brightness, falling into the shadow of the earth, was imagined by the ancients to sicken and swoon, as if she were going to die. By some very ancient nations she was supposed, at such times, to be in pain; and, in order to relieve her fancied distress, they lifted torches high in the atmosphere, blew horns and trumpets, beat upon brazen vessels, and even, after the eclipse was over, they offered sacrifices to the moon. The opinion also extensively prevailed, that it was in the power of witches, by their spells and charms, not only to darken the moon, but to bring her down from her orbit, and to compel her to shed her baleful influences upon the earth. In solar eclipses, also, especially when total, the sun was supposed to turn away his face in abhorrence of some atrocious crime, that either had been perpetrated or was about to be perpetrated, and to threaten mankind with everlasting night, and the destruction of the world. To such superstitions Milton alludes, in the passage which I have taken for the motto of this Letter.

The Chinese, who, from a very high period of antiquity, have been great observers of eclipses, although they did not take much notice of those of the moon, regarded eclipses of the sun in general as unfortunate, but especially such as occurred on the first day of the year. These were thought to forebode the greatest calamities to the emperor, who on such occasions did not receive the usual compliments of the season. When, from the predictions of their astronomers, an eclipse of the sun was expected, they made great preparation at court for observing it; and as soon as it commenced, a blind man beat a drum, a great concourse assembled, and the mandarins, or nobility, appeared in state.

LETTER XIX.

LONGITUDE.—TIDES.

"First in his east, the glorious lamp was seen,
Regent of day, and all the horizon round
Invested with bright rays, jocund to run
His *longitude* through heaven's high road; the gray and the
Pleiades before him danced, Shedding sweet influence."—*Milton.*

THE ancients studied astronomy chiefly as subsidiary to astrology, with the vain hope of thus penetrating the veil of futurity, and reading their destinies among the stars. The moderns, on the other hand, have in view, as the great practical object of this study, the perfecting of the art of navigation. When we reflect on the vast interests embarked on the ocean, both of property and life, and upon the immense benefits that accrue to society from a safe and speedy intercourse between the different nations of the earth, we cannot but see that whatever tends to enable the mariner to find his way on the pathless ocean, and to secure him against its multiplied dangers, must confer a signal benefit on society.

In ancient times, to venture out of sight of land was deemed an act of extreme audacity; and Horace, the Roman poet, pronounces him who first ventured to trust his frail bark to the stormy ocean, endued with a heart of oak, and girt with triple folds of brass. But now, the navigator who fully avails himself of all the resources of science, and especially of astronomy, may launch fearlessly on the deep, and almost bid defiance to rocks and tempests. By enabling the navigator to find his place on the ocean with almost absolute precision, however he may have been driven about by the winds, and however long he may have been out of sight of land, astronomers must be held as great benefactors to all who commit either their lives or their fortunes to the sea. Nor have they secured to the art of navigation such benefits without incredible study and toil, in watching the motions of the heavenly bodies, in investigating the laws by which their movements are governed, and in reducing all their discoveries to a form easily available to the navigator, so that, by some simple observation on one or two of the heavenly bodies, with instruments which the astronomer has invented, and prepared for his use, and by looking out a few numbers in tables which have been compiled for him, with immense labor, he may ascertain the exact place he occupies on the surface of the globe, thousands of miles from land.

The situation of any place is known by its latitude and longitude. As charts of every ocean and sea are furnished to the sailor, in which are laid down the latitudes and longitudes of every point of land, whether on the shores of islands or the main, he has, therefore, only to ascertain his latitude and longitude at any particular place on the ocean, in order to find where he is, with respect to the nearest point of land, although this may be, and may always have been, entirely out of sight to him.

To determine the *latitude* of a place is comparatively an easy matter, whenever we can see either the sun or the stars. The distance of the sun from the zenith, when on the meridian, on a given day of the year, (which distance we may easily take with the sextant,) enables us, with the aid of the tables, to find the latitude of the place; or, by taking the altitude of the north star, we at once obtain the latitude.

The *longitude* of a place may be found by any method, by which we may ascertain how much its time of day differs from that of Greenwich at the same moment. A place that lies eastward of another comes to the meridian an hour earlier for every fifteen degrees of longitude, and of course has the hour of the day so much in advance of the other, so that it counts one o'clock when the other place counts twelve. On the other hand, a place lying westward of another comes to the meridian later by one hour for every fifteen degrees, so that it counts only eleven o'clock when the other place counts twelve. Keeping these principles in view, it is easy to see that a comparison of the difference of time between two places at the same moment, allowing fifteen degrees for an hour, sixty minutes for every four minutes of time, and sixty seconds for every four seconds of time, affords us an accurate mode of finding the difference of longitude between the two places. This comparison may be made by means of a chronometer, or from solar or lunar eclipses, or by what is called the lunar method of finding the longitude.

Chronometers are distinguished from clocks, by being regulated by means of a balance-wheel instead of a pendulum. A watch, therefore, comes under the general definition of a chronometer; but the name is more commonly applied to larger timepieces, too large to be carried about the person, and constructed with the greatest possible accuracy, with special reference to finding the longitude. Suppose, then, we are furnished with a chronometer set to Greenwich time. We arrive at New York, for example, and compare it with the time there. We find it is five hours in advance of the New-York time, indicating five o'clock, P.M., when it is noon at New York. Hence we find that the longitude of New York is 5×15=75 degrees.[11] The time at New York, or any individual place, can be known by observations with the transit-instrument, which gives us the precise moment when the sun is on the meridian.

It would not be necessary to resort to Greenwich, for the purpose of setting our chronometer to Greenwich time, as it might be set at any place whose longitude is known, having been previously determined. Thus, if we know that the longitude of a certain place is exactly sixty degrees east of Greenwich, we have only to set our chronometer four hours behind the time at that place, and it will be regulated to Greenwich time. Hence it is a matter of the greatest importance to navigation, that the longitude of numerous ports, in different parts of the earth, should be accurately determined, so that when a ship arrives at any such port, it may have the means of setting or verifying its chronometer.

This method of taking the longitude seems so easy, that you will perhaps ask, why it is not sufficient for all purposes, and accordingly, why it does not supersede the move complicated and laborious methods? why every sailor does not provide himself with a chronometer, instead of finding his longitude at sea by tedious and oft-repeated calculations, as he is in the habit of doing? I answer, it is only in a few extraordinary cases that chronometers have been constructed of such accuracy as to afford results as exact as those obtained by the other methods, to be described shortly; and instruments of such perfection are too expensive for general use among sailors. Indeed, the more common chronometers cost too much to come within the means of a great majority of sea-faring men. Moreover, by being transported from place to place, chronometers are liable to change their *rate*. By the rate of any timepiece is meant its deviation from perfect accuracy. Thus, if a clock should gain one second per day, one day with another, and we should find it impossible to bring it nearer to the truth, we may reckon this as its rate, and allow for it in our estimate of the time of any particular observation. If the error was not uniform, but sometimes greater and sometimes less than one second per day, then the amount of such deviation is called its "variation from its mean rate." I introduce these minute statements, (which are more precise than I usually deem necessary,) to show you to what an astonishing degree of accuracy chronometers have in some instances been brought. They have been carried from London to Baffin's Bay, and brought back, after a three years' voyage, and found to have varied from their mean rate, during the whole time, only a second or two, while the extreme variation of several chronometers, tried at the Royal Observatory at Greenwich, never exceeded a second and a half. Could chronometers always be depended on to such a degree of accuracy as this, we should hardly desire any thing better for determining the longitude of different places on the earth. A recent determination of the longitude of the City Hall in New York, by means of three chronometers, sent out from

London expressly for that purpose, did not differ from the longitude as found by a solar eclipse (which is one of the best methods) but a second and a quarter.

Eclipses of the sun and moon furnish the means of ascertaining the longitude of a place, because the entrance of the moon into the earth's shadow in a lunar eclipse, and the entrance of the moon upon the disk of the sun in a solar eclipse, are severally examples of one of those instantaneous occurrences in the heavens, which afford the means of comparing the times of different places, and of thus determining their differences of longitude. Thus, if the commencement of a lunar eclipse was seen at one place an hour sooner than at another, the two places would be fifteen degrees apart, in longitude; and if the longitude of one of the places was known, that of the other would become known also. The exact instant of the moon's entering into the shadow of the earth, however, cannot be determined with very great precision, since the moon, in passing through the earth's penumbra, loses its light gradually, so that the moment when it leaves the penumbra and enters into the shadow cannot be very accurately defined. The first contact of the moon with the sun's disk, in a solar eclipse, or the moment of leaving it,—that is, the beginning and end of the eclipse,—are instants that can be determined with much precision, and accordingly they are much relied on for an accurate determination of the longitude. But, on account of the complicated and laborious nature of the calculation of the longitude from an eclipse of the sun, (since the beginning and end are not seen at different places, at the same moment,) this method of finding the longitude is not adapted to common use, nor available at sea. It is useful, however, for determining the longitude of fixed observatories. The *lunar method of finding the longitude* is the most refined and accurate of all the modes practised at sea. The motion of the moon through the heavens is so rapid, that she perceptibly alters her distance from any star every minute; consequently, the moment when that distance is a certain number of degrees and minutes is one of those instantaneous events, which may be taken advantage of for comparing the times of different places, and thus determining their difference of longitude. Now, in a work called the 'Nautical Almanac,' printed in London, annually, for the use of navigators, the distance of the moon from the sun by day, or from known fixed stars by night, for every day and night in the year, is calculated beforehand. If, therefore, a sailor wishes to ascertain his longitude, he may take with his sextant the distance of the moon from one of these stars at any time,— suppose nine o'clock, at night,—and then turn to the 'Nautical Almanac,' and see *what time it was at Greenwich* when the distance between the moon and that star was the same. Let it be twelve o'clock, or three hours in advance of his time: his longitude, of course, is forty-five degrees west.

This method requires more skill and accuracy than are possessed by the majority of seafaring men; but, when practised with the requisite degree of skill, its results are very satisfactory. Captain Basil Hall, one of the most scientific commanders in the British navy, relates the following incident, to show the excellence of this method. He sailed from San Blas, on the west coast of Mexico, and, after a voyage of eight thousand miles, occupying eighty-nine days, arrived off Rio de Janeiro, having, in this interval, passed through the Pacific Ocean, rounded Cape Horn, and crossed the South Atlantic, without making any land, or even seeing a single sail, with the exception of an American whaler off Cape Horn. When within a week's sail of Rio, he set seriously about determining, by lunar observations, the precise line of the ship's course, and its situation at a determinate moment; and having ascertained this within from five to ten miles, ran the rest of the way by those more ready and compendious methods, known to navigators, which can be safely employed for short trips between one known point and another, but which cannot be trusted in long voyages, where the moon is the only sure guide. They steered towards Rio Janeiro for some days after taking the lunars, and, having arrived within fifteen or twenty miles of the coast, they hove to, at four in the morning, till the day should break, and then bore up, proceeding cautiously, on account of a thick fog which enveloped them. As this cleared away, they had the satisfaction of seeing the great Sugar-Loaf Rock, which stands on one side of the harbor's mouth, so nearly right ahead, that they had not to alter their course above a point, in order to hit the entrance of the harbor. This was the first land they had seen for three months, after crossing so many seas, and being set backwards and forwards by innumerable currents and foul winds. The effect on all on board was electric; and the admiration of the sailors was unbounded. Indeed, what could be more admirable than that a man on the deck of a vessel, by measuring the distance between the moon and a star, with a little instrument which he held in his hand, could determine his exact place on the earth's surface in the midst of a vast ocean, after having traversed it in all directions, for three months, crossing his track many times, and all the while out of sight of land?

The lunar method of finding the longitude could never have been susceptible of sufficient accuracy, had not the motions of the moon, with all their irregularities, been studied and investigated by the most laborious and profound researches. Hence Newton, while wrapt in those meditations which, to superficial minds, would perhaps have appeared rather curious than useful, inasmuch as they respected distant bodies of the universe which seemed to have little connexion with the affairs of this world, was laboring night and day for the benefit of the sailor and the merchant. He was guiding the vessel of the one, and securing the merchandise of the other; and thus he contributed a large share to promote the happiness of his fellow-men, not only in exalting the powers of the human intellect, but also in preserving the lives and fortunes of those engaged in navigation and commerce. Principles in science are rules in art; and the philosopher who is engaged in the investigation of these principles, although his pursuits may be thought less practically useful than those of the artisan who carries out those principles into real life, yet, without the knowledge of the principles, the rules would have never been known. Studies, therefore, the most abstruse, are, when viewed as furnishing rules to act, often productive of the highest practical utility.

Since the *tides* are occasioned by the influence of the sun and moon, I will conclude this Letter with a few remarks on this curious phenomenon. By the tides are meant the alternate rising and falling of the waters of the ocean. Its greatest and least elevations are called *high and low water*; its rising and falling are called *flood and ebb*; and the extraordinary high and low tides that occur twice every month are called *spring and neap tides*. It is high or low tide on opposite sides of the globe at the same time. If, for example, we have high water at noon, it is also high water to those who live on the meridian below us, where it is midnight. In like manner, low water occurs simultaneously on opposite sides of the meridian. The average amount of the tides for the whole globe is about two and a half feet; but their actual height at different places is very various, sometimes being scarcely perceptible, and sometimes rising to sixty or seventy feet. At the same place, also, the phenomena of the tides are very different at different times. In the Bay of Fundy, where the tide rises seventy feet, it comes in a mighty wave, seen thirty miles off, and roaring with a loud noise. At the mouth of the Severn, in England, the flood comes up in one head about ten feet high, bringing certain destruction to any small craft that has been unfortunately left by the ebbing waters on the flats and as it passes the mouth of the Avon, it sends up that small river a vast body of water, rising, at Bristol, forty or fifty feet.

Tides are caused by the unequal attractions of the sun and moon upon different parts of the earth. Suppose the projectile force by which the earth is carried forward in her orbit to be suspended, and the earth to fall towards one of these bodies,—the moon, for example,—in consequence of their mutual attraction. Then, if all parts of the earth fell equally towards the moon, no derangement of its different parts would result, any more than of the particles of a drop of water, in its descent to the ground. But if one part fell faster than another, the different portions would evidently be separated from each other. Now, this is precisely what takes place with respect to the earth, in its fall towards the moon. The portions of the earth in the hemisphere next to the moon, on account of being nearer to the centre of attraction, fall faster than those in the opposite hemisphere, and consequently leave them behind. The solid earth, on account of its cohesion, cannot obey this impulse, since all its different portions constitute one mass, which is acted on in the same manner as though it were all collected in the centre; but the waters on the surface, moving freely under this impulse, endeavor to desert the solid mass and fall towards the moon. For a similar reason, the waters in the opposite hemisphere, falling less towards the moon than the solid earth does, are left behind, or appear to rise.

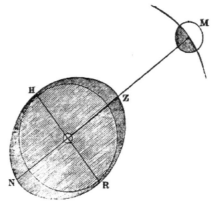

Fig. 46.

But if the moon draws the waters of the earth into an oval form towards herself, raising them simultaneously on the opposite sides of the earth, they must obviously be drawn away from the intermediate parts of the earth, where it must at the same time be low water. Thus, in Fig. 46, the moon, M, raises the waters beneath itself at Z and N, at which places it is high water, but at the same time depresses the waters at H and R, at which places it is low water. Hence, the interval between the high and low tide, on successive days, is about fifty minutes, corresponding to the progress of the moon in her orbit from west to east, which causes her to come to the meridian about fifty minutes later every day. There occurs, however, an intermediate tide, when the moon is on the lower meridian, so that the interval between two high tides is about twelve hours, and twenty-five minutes.

Were it not for the impediments which prevent the force from producing its full effects, we might expect to see the great tide-wave, as the elevated crest is called, always directly beneath the moon, attending it regularly around the globe. But the inertia of the waters prevents their instantly obeying the moon's attraction, and the friction of the waters on the bottom of the ocean still further retards its progress. It is not, therefore, until several hours (differing at different places) after the moon has passed the meridian of a place, that it is high tide at that place.

The *sun* has an action similar to that of the moon, but only *one third* as great. On account of the great mass of the sun, compared with that of the moon, we might suppose that his action in raising the tides would be greater than the moon's; but the nearness of the moon to the earth more than compensates for the sun's greater quantity of matter. As, however, wrong views are frequently entertained on this subject, let us endeavor to form a correct idea of the advantage which the moon derives from her proximity. It is not that her actual amount of attraction is thus rendered greater than that of the sun; but it is that her attraction for the *different parts* of the earth is very unequal, while that of the sun is nearly uniform. It is the *inequality* of this action, and not the absolute force, that produces the tides. The sun being ninety-five millions of miles from the earth, while the diameter of the earth is only one twelve thousandth part of this distance, the effects of the sun's attraction will be nearly the same on all parts of the earth, and therefore will not, as was explained of the moon, tend to separate the waters from the earth on the nearest side, or the earth from the waters on the remotest side, but in a degree proportionally smaller. But the diameter of the earth is one thirtieth the distance of the moon, and therefore the moon acts with considerably greater power on one part of the earth than on another.

As the sun and moon both contribute to produce the tides, and as they sometimes act together and sometimes in opposition to each other, so corresponding variations occur in the height of the tide. The *spring tides*, or those which rise to an unusual height twice a month, are produced by the sun and moon's acting together; and the *neap tides*, or those which are unusually low twice a month, are produced by the sun and moon's acting in opposition to each other. The spring tides occur at the syzygies: the neap tides at the quadratures. At the time of new moon, the sun and moon both being on the same side of the earth, and acting upon it in the same line, their actions conspire, and the sun may be considered as adding so much to the force of the moon. We have already seen how the moon contributes to raise a tide on the opposite side of the earth. But the sun, as well as the moon, raises its own tide-wave, which at new moon coincides with the lunar tide-wave. This will be plain on inspecting the diagram, Fig. 47, on page 220, where S represents the sun, C, the moon in conjunction, O, the moon in opposition, and Z, N, the tide-wave. Since the sun and moon severally raise a tide-wave, and the two here coincide, it is evident that a peculiarly high tide must occur when the two bodies are in conjunction, or at new moon. At full moon, also, the two luminaries conspire in the same way to raise the tide; for we must recollect that each body contributes to raise a tide on the opposite side. Thus, when the sun is at S and the moon at O, the sun draws the waters on the side next to it away from the earth, and the moon draws the earth away from the waters on that side; their united actions, therefore, conspire, and an unusually high tide is the result. On the side next to O, the two forces likewise conspire: for while the moon draws the waters away from the earth, the sun draws the earth away from the waters. In both cases an unusually low tide is produced; for the more the water is elevated at Z and N, the more it will be depressed at H and R, the places of low tide.

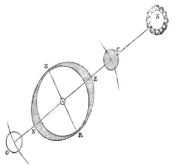

Fig. 47.

Twice a month, also, namely, at the quadratures of the moon, the tides neither rise so high nor fall so low as at other times, because then the sun and moon act against each other. Thus, in Fig. 48, while F tends to raise the water at Z, S tends to depress it, and consequently the high tide is less than usual. Again, while F tends to depress the water at R, S tends to elevate it, and therefore the low tide is less than usual. Hence the difference between high and low water is only half as great at neap as at spring tide. In the diagrams, the elevation and depression of the waters is represented, for the sake of illustration, as far greater than it really is; for you must recollect that the average height of the tides for the whole globe is only about two and a half feet, a quantity so small, in comparison with the diameter of the earth, that were the due proportions preserved in the figures, the effect would be wholly insensible.

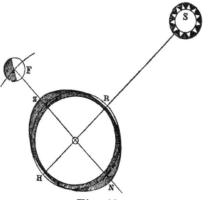

Fig. 48.

The variations of distance in the sun are not great enough to influence the tides very materially, but the variations in the moon's distances have a striking effect. The tides which happen, when the moon is in perigee, are considerably greater than when she is in apogee; and if she happens to be in perigee at the time of the syzygies, the spring tides are unusually high.

The motion of the tide-wave is not a *progressive* motion, but a mere undulation, and is to be carefully distinguished from the currents to which it gives rise. If the ocean completely covered the earth, the sun and moon being in the equator, the tide-wave would travel at the same rate as the earth revolves on its axis. Indeed, the correct way of conceiving of the tide-wave, is to consider the moon at rest, and the earth, in its rotation from west to east, as bringing successive portions of water under the moon, which portions being elevated successively, at the same rate as the earth revolves on its axis, have a relative motion westward, at the same rate.

The tides of rivers, narrow bays, and shores far from the main body of the ocean, are not produced in those places by the direct action of the sun and moon, but are subordinate waves propagated from the great tide-wave, and are called *derivative* tides, while those raised directly by the sun and moon are called *primitive* tides.

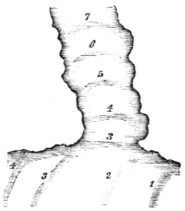

Fig. 49.

The velocity with which the tide moves will depend on various circumstances, but principally on the depth, and probably on the regularity, of the channel. If the depth is nearly uniform the tides will be regular; but if some parts of the channel are deep while others are shallow, the waters will be detained by the greater friction of the shallow places, and the tides will be irregular. The direction, also, of the derivative tide may be totally different from that of the primitive. Thus, in Fig. 49, if the great tide-wave, moving from east to west, is represented by the lines 1, 2, 3, 4, the derivative tide, which is propagated up a river or bay, will be represented by the lines 3, 4, 5, 6, 7. Advancing faster in the channel than next the bank, the tides will lag behind towards the shores, and the tide-wave will take the form of curves, as represented in the diagram.

On account of the retarding influence of shoals, and an uneven, indented coast, the tide-wave travels more slowly along the shores of an island than in the neighboring sea, assuming convex figures at a little distance from the island, and on opposite sides of it. These convex lines sometimes meet, and become blended in such a way, as to create singular anomalies in a sea much broken by islands, as well as on coasts indented with numerous bays and rivers. Peculiar phenomena are also produced, when the tide flows in at opposite extremities of a reef or island, as into the two opposite ends of Long-Island Sound. In certain cases, a tide-wave is forced into a narrow arm of the sea, and produces very remarkable tides. The tides of the Bay of Fundy (the highest in the world) are ascribed to this cause. The tides on the coast of North America are derived from the great tide-wave of the South Atlantic, which runs steadily northward along the coast to the mouth of the Bay of Fundy, where it meets the northern tide-wave flowing in the opposite direction. This accumulated wave being forced into the narrow space occupied by the Bay, produces the remarkable tide of that place.

The largest lakes and inland seas have no perceptible tides. This is asserted by all writers respecting the Caspian and Euxine; and the same is found to be true of the largest of the North-American lakes, Lake Superior. Although these several tracts of water appear large, when taken by themselves, yet they occupy but small portions of the surface of the globe, as will appear evident from the delineation of them on the artificial globe. Now, we must recollect that the primitive tides are produced by the *unequal* action of the sun and moon upon the different parts of the earth; and that it is only at points whose distance from each other bears a considerable ratio to the whole distance of the sun or moon, that the inequality of action becomes manifest. The space required to make the effect sensible is larger than either of these tracts of water. It is obvious, also, that they have no opportunity to be subject to a derivative tide.

Although all must admit that the tides have *some connexion* with the sun and the moon, yet there are so many seeming anomalies, which at first appear irreconcilable with the theory of gravitation, that some are unwilling to admit the explanation given by this theory. Thus, the height of the tide is so various, that at some places on the earth there is scarcely any tide at all, while at other places it rises to seventy feet. The time of occurrence is also at many places wholly unconformable to the motions of the moon, as is required by the theory, being low water where it should be high water; or, instead of appearing just beneath the moon, as the theory would lead us to expect, following it at the distance of six, ten, or even fifteen, hours; and finally, the moon sometimes appears to have no part at all in producing the tide, but it happens uniformly at noon and midnight, (as is said to be the case at the Society Islands,) and therefore seems wholly dependent on the sun.

Notwithstanding these seeming inconsistencies with the law of universal gravitation, to which the explanation of the tides is referred, the correspondence of the tides to the motions of the sun and moon, in obedience to the law of attraction, is in general such as to warrant the application of that law to them, while in a great majority of the cases which appear to be exceptions to the operation of that law, local causes and impediments have been discovered, which modified or overruled the uniform operation of the law of gravitation. Thus it does not disprove the reality of the existence of a force which carries bodies near the surface of the earth towards its centre, that we see them sometimes compelled, by the operation of local causes, to move in the opposite direction. A ball shot from a cannon is still subject to the law of gravitation, although, for a certain time, in obedience to the impulse given it, it may proceed in a line contrary to that in which gravity alone would carry it. The fact that water may be made to run up hill does not disprove the fact that it usually runs down hill by the force of gravity, or that it is still subject to this force, although, from the action of modifying or superior forces, it may be proceeding in a direction contrary to that given by gravity. Indeed, those who have studied the doctrine of the tides most profoundly consider them as affording a striking and palpable exhibition of the truth of the doctrine of universal gravitation.

LETTER XX.

PLANETS.—MERCURY AND VENUS.

"First, Mercury, amidst full tides of light,
Rolls next the sun, through his small circle bright;
Our earth would blaze beneath so fierce a ray,
And all its marble mountains melt away.
Fair Venus next fulfils her larger round,
With softer beams, and milder glory crowned;
Friend to mankind, she glitters from afar,
Now the bright evening, now the morning, star."—*Baker.*

THERE is no study in which more is to be hoped for from a lucid arrangement, than in the study of astronomy. Some subjects involved in this study appear very difficult and perplexing to the learner, before he has fully learned the doctrine of the sphere, and gained a certain familiarity with astronomical doctrines, which would seem very easy to him after he had made such attainments. Such an order ought to be observed, as shall bring out the facts and doctrines of the science just in the place where the mind of the learner is prepared to receive them. Some writers on astronomy introduce their readers at once to the most perplexing part of the whole subject,—the planetary motions. I have thought a different course advisable, and have therefore commenced these Letters with an account of those bodies which are most familiarly known to us, the earth, the sun, and the moon. In connexion with the earth, we are able to acquire a good knowledge of the artificial divisions and points of reference that are established on the earth and in the heavens, constituting the doctrine of the sphere. You thus became familiar with many terms and definitions which are used in astronomy. These ought to be always very clearly borne in mind; and if you now meet with any term, the definition of which you have either partially or wholly forgotten, let me strongly recommend to you, to turn back and review it, until it becomes as familiar to you as household words. Indeed, you will find it much to your advantage to go back frequently, and reiterate the earlier parts of the subject, before you advance to subjects of a more intricate nature. If this process should appear to you a little tedious, still you will find yourself fully compensated by the clear light in which all the succeeding subjects will appear. This clear and distinct perception of the ground we have been over shows us just where we are on our journey, and helps us to find the remainder of the way with far greater ease than we could otherwise do. I do not, however, propose by any devices to relieve you from the trouble of thinking. Those who are not willing to incur this trouble can never learn much of astronomy.

In introducing you to the planets, (which next claim our attention,) I will, in the first place, endeavor to convey to you some clear views of these bodies individually, and afterwards help you to form as correct a notion as possible of their motions and mutual relations.

The name *planet* is derived from a Greek word, (πλανητης, *planetes*,) which signifies a *wanderer*, and is applied to this class of bodies, because they shift their positions in the heavens, whereas the fixed stars constantly maintain the same places with respect to each other. The planets known from a high antiquity are, Mercury, Venus, Earth, Mars, Jupiter, and Saturn. To these, in 1781, was added Uranus, (or *Herschel*, as it is sometimes called, from the name of its discoverer;) and, as late as the commencement of the present century, four more were added, namely, Ceres, Pallas, Juno, and Vesta. These bodies are designated by the following characters:

1. Mercury, ☿	7. Ceres, ?
2. Venus, ♀	8. Pallas, ⚴
3. Earth, ♁	9. Jupiter, ♃
4. Mars, ♂	10. Saturn, ♄
5. Vesta, ⚶	11. Uranus, ♅
6. Juno, ⚵	

The foregoing are called the *primary* planets. Several of these have one or more attendants, or satellites, which revolve around them as they revolve around the sun. The Earth has one satellite, namely, the Moon; Jupiter has four; Saturn, seven; and Uranus, six. These bodies are also planets, but, in distinction from the others, they are called *secondary* planets. Hence, the whole number of planets are twenty-nine, namely, eleven primary, and eighteen secondary, planets.

You need never look for a planet either very far in the north or very far in the south, since they are always near the ecliptic. Mercury, which deviates furthest from that great circle, never is seen more than seven degrees from it; and you will hardly ever see one of the planets so far from it as this, but they all pursue nearly the same great route through the skies, in their revolutions around the sun. The new planets, however, make wider excursions from the plane of the ecliptic, amounting, in the case of Pallas, to thirty-four and a half degrees.

Mercury and Venus are called *inferior* planets, because they have their orbits nearer to the sun than that of the earth; while all the others, being more distant from the sun than the earth, are called *superior* planets. The planets present great diversities among themselves, in respect to distance from the sun, magnitude, time of revolution, and density. They differ, also, in regard to satellites, of which, as we have seen, three have respectively four, six, and seven, while more than half have none at all. It will aid the memory, and render our view of the planetary system more clear and comprehensive, if we classify, as far as possible, the various particulars comprehended under the foregoing heads. As you have had an opportunity, in preceding Letters, of learning something respecting the means which astronomers have of ascertaining the distances and magnitudes of these bodies, you will not doubt that they are really as great as they are represented; but when you attempt to conceive of spaces so vast, you will find the mind wholly inadequate to the task. It is indeed but a comparatively small space that we can fully comprehend at one grasp. Still, by continual and repeated efforts, we may, from time to time, somewhat enlarge the boundaries of our mental vision. Let us begin with some known and familiar space, as the distance between two places we are accustomed to traverse. Suppose this to be one hundred miles. Taking this as our measure, let us apply it to some greater distance, as that across the Atlantic Ocean,—say three thousand miles. From this step we may advance to some faint conception of the diameter of the earth; and taking that as a new measure, we may apply it to such greater spaces as the distance of the planets from the sun. I hope you will make trial of this method on the following comparative statements respecting the planets.

Distances from the Sun, in miles.

1. Mercury.	37,000,000	6. Juno,	}	
2. Venus,	68,000,000	7. Ceres,	}	261,000,000
3. Earth.	95,000,000	8. Pallas,	}	
4. Mars,	142,000,000	9. Jupiter.	485,000,000	
5. Vesta,	225,000,000	10. Saturn.	890,000,000	
		11. Uranus, or Herschel,	1800,000,000	

The *dimensions* of the planetary system are seen from this table to be vast, comprehending a circular space thirty-six hundred millions of miles in diameter. A rail-way car, travelling constantly at the rate of twenty miles an hour, would require more than twenty thousand years to cross the orbit of Uranus.

	Diam. in miles.			Diam. in miles.
1. Mercury,	3140		5. Ceres,	160
2. Venus,	7700		6. Jupiter,	89,000
3. Earth,	7912		7. Saturn,	79,000
4. Mars,	4200		8. Uranus,	35,000

We remark here a great diversity in regard to magnitude,—a diversity which does not appear to be subject to any definite law. While Venus, an inferior planet, is nine tenths as large as the earth, Mars, a superior planet, is only one seventh, while Jupiter is twelve hundred and eighty-one times as large. Although several of the planets, when nearest to us, appear brilliant and large, when compared with most of the fixed stars, yet the angle which they subtend is very small,—that of Venus, the greatest of all, never exceeding about one minute, which is less than one thirtieth the apparent diameter of the sun or moon. Jupiter, also, by his superior brightness, sometimes makes a striking figure among the stars; yet his greatest apparent diameter is less than one fortieth that of the sun.

Periodic Times.

Mercury	revolves	around	the sun	in nearly	3	months.
Venus,	"	"	"	"	7½	"
Earth,	"	"	"	"	1	year.
Mars,	"	"	"	"	2	years.
Ceres,	"	"	"	"	4½ y.	"
Jupiter,	"	"	"	"	12	"
Saturn,	"	"	"	"	29	"
Uranus,	"	"	"	"	84	"

From this view, it appears that the planets nearest the sun move most rapidly. Thus, Mercury performs nearly three hundred and fifty revolutions while Uranus performs one. The apparent progress of the most distant planets around the sun is exceedingly slow. Uranus advances only a little more than four degrees in a whole year; so that we find this planet occupying the same sign, and of course remaining nearly in the same part of the heavens, for several years in succession.

After this comparative view of the planets in general, let us now look at them individually; and first, of the inferior planets, Mercury and Venus.

MERCURY and VENUS, having their orbits so far within that of the earth, appear to us as attendants upon the sun. Mercury never appears further from the sun than twenty-nine degrees, and seldom so far; and Venus, never more than about forty-seven degrees. Both planets, therefore, appear either in the west soon after sunset, or in the east a little before sunrise. In high latitudes, where the twilight is long, Mercury can seldom be seen with the naked eye, and then only when its angular distance from the sun is greatest. Copernicus, the great Prussian astronomer, (who first distinctly established the order of the solar system, as at present received,) lamented, on his death-bed, that he had never been able to obtain a sight of Mercury; and Delambre, a distinguished astronomer of France, saw it but twice. In our latitude, however, we may see this planet for several evenings and mornings, if we will watch the time (as usually given in the almanac) when it is at its greatest elongations from the sun. It will not, however, remain long for our gaze, but will soon run back to the sun. The reason of this will be readily understood from the following diagram, Fig. 50. Let S represent the sun, E, the earth, and M, N, Mercury at its greatest elongations from the sun, and O Z P, a portion of the sky. Then, since we refer all distant bodies to the same concave sphere of the heavens, it is evident that we should see the sun at Z, and Mercury at O, when at its greatest eastern elongation, and at P, when at its greatest western elongation; and while passing from M to N through Q, it would appear to describe the arc O P; and while passing from N to M through R, it would appear to run back across the sun on the same arc. It is further evident that it would be visible only when at or near one of its greatest elongations; being at all other times so near the sun as to be lost in his light.

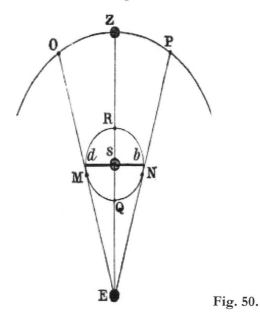

Fig. 50.

A planet is said to be in *conjunction* with the sun when it is seen in the same part of the heavens with the sun. Mercury and Venus have each two conjunctions, the inferior and the superior conjunction. The *inferior conjunction* is its position when in conjunction on the same side of the sun with the earth, as at Q, in the figure; the *superior conjunction* is its position when on the side of the sun most distant from the earth, as at R.

The time which a planet occupies in making one entire circuit of the heavens, from any star, until it comes round to the same star again, is called its *sidereal revolution.* The period occupied by a planet between two successive conjunctions with the earth is called its *synodical revolution.* Both the planet and the earth being in motion, the time of the synodical revolution of Mercury or Venus exceeds that of the sidereal; for when the planet comes round to the place where it before overtook the earth, it does not find the earth at that point, but far in advance of it. Thus, let Mercury come into inferior conjunction with the earth at C, Fig. 51. In about eighty-eight days, the planet will come round to the same point again; but, mean-while, the earth has moved forward through the arc E E´, and will continue to move while the planet is moving more rapidly to overtake her; the case being analogous to that of the hour and minute hand of a clock.

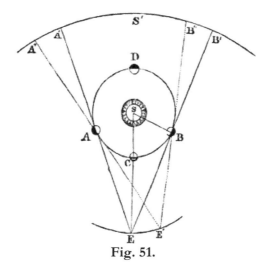

Fig. 51.

The synodical period of Mercury is one hundred and sixteen days, and that of Venus five hundred and eighty-four days. The former is increased twenty-eight days, and the latter, three hundred and sixty days, by the motion of the earth; so that Venus, after being in conjunction with the earth, goes more than twice round the sun before she comes into conjunction again. For, since the earth is likewise in motion, and moves more than half as fast as Venus, by the time the latter has gone round and returned to the place where the two bodies were together, the earth is more than half way round, and continues moving, so that it will be a long time before Venus comes up with it.

The motion of an inferior planet is *direct* in passing through its superior conjunction, and *retrograde* in passing through its inferior conjunction. You will recollect that the motion of a heavenly body is said to be direct when it is in the order of the signs from west to east, and retrograde when it is contrary to the order of the signs, or from east to west. Now Venus, while going from B through D to A, (Fig. 51,) moves from west to east, and would appear to traverse the celestial vault B′ S′ A′, from right to left; but in passing from A through C to B, her course would be retrograde, returning on the same arc from left to right. If the earth were at rest, therefore, (and the sun, of course, at rest,) the inferior planets would appear to oscillate backwards and forwards across the sun. But it must be recollected that the earth is moving in the same direction with the planet, as respects the signs, but with a slower motion. This modifies the motions of the planet, accelerating it in the superior, and retarding it in the inferior, conjunction. Thus, in Fig. 51, Venus, while moving through B D A, would seem to move in the heavens from B′ to A′, were the earth at rest; but, mean-while, the earth changes its position from E to E′, on which account the planet is not seen at A′, but at A″, being accelerated by the arc A′ A″, in consequence of the earth's motion. On the other hand, when the planet is passing through its inferior conjunction A C B, it appears to move backwards in the heavens from A′ to B′, if the earth is at rest, but from A′ to B″, if the earth has in the mean time moved from E to E′, being retarded by the arc B′ B″. Although the motions of the earth have the effect to accelerate the planet in the superior conjunction, and to retard it in the inferior, yet, on account of the greater distance, the apparent motion of the planet is much slower in the superior than in the inferior conjunction, Venus being the whole breadth of her orbit, or one hundred and thirty-six millions of miles further from us when at her greatest, than when at her least, distance, as is evident from Fig. 51. When passing from the superior to the inferior conjunction, or from the inferior to the superior, through the greatest elongations, the inferior planets are *stationary*. Thus, (Fig. 51,) when the planet is at A, the earth being at E, as the planet's motion is directly towards the spectator, he would constantly project it at the same point in the heavens, namely, A′; consequently, it would appear to stand still. Or, when at its greatest elongation on the other side, at B, as its motion would be directly from the spectator, it would be seen constantly at B′. If the earth were at rest, the stationary points would be at the greatest elongations, as at A and B; but the earth itself is moving nearly at right angles to the planet's motion, which makes the planet appear to move in the opposite direction. Its direct motion will therefore continue longer on the one side, and its retrograde motion longer on the other side, than would be the case, were it not for the motion of the earth. Mercury, whose greatest angular distance from the sun is nearly twenty-nine degrees, is stationary at an elongation of from fifteen to twenty degrees; and Venus, at about twenty-nine degrees, although her greatest elongation is about forty-seven degrees.

Mercury and Venus exhibit to the telescope *phases* similar to those of the moon. When on the side of their inferior conjunction, as from B to C through D, Fig. 52, less than half their enlightened disk is turned towards us, and they appear horned, like the moon in her first and last quarters; and when on the side of the superior conjunction, as from C to B through A, more than half the enlightened disk is turned towards us, and they appear gibbous. At the moment of superior conjunction, the whole enlightened orb of the planet is turned towards the earth, and the appearance would be that of the full moon; but the planet is too near the sun to be commonly visible.

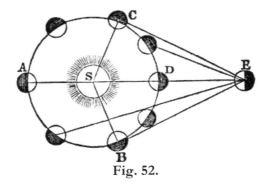

Fig. 52.

We should at first thought expect, that each of these planets would be largest and brightest near their inferior conjunction, being then so much nearer to us than at other times; but we must recollect that, when in this situation, only a small part of the enlightened disk is turned toward us. Still, the period of greatest brilliancy cannot be when most of the illuminated side is turned towards us, for then, being at the superior conjunction, its light will be diminished, both by its great distance, and by its being so near the sun as to be partially lost in the twilight. Hence, when Venus is a little within her place of greatest elongation, about forty degrees from the sun, although less than half her disk is enlightened, yet, being comparatively near to us, and shining at a considerable altitude after the evening or before the morning twilight, she then appears in greatest splendor, and presents an object admired for its beauty in all ages. Thus Milton,

"Fairest of stars, last in the train of night, If better thou belong not to the dawn, Sure pledge of day that crown'st the smiling morn With thy bright circlet."

Mercury and Venus both *revolve on their axes* in nearly the same time with the earth. The diurnal period of Mercury is a little greater, and that of Venus a little less, than twenty-four hours. These revolutions have been determined by means of some spot or mark seen by the telescope, as the revolution of the sun on his axis is ascertained by means of his spots. Mercury owes most of its peculiarities to its proximity to the sun. Its light and heat, derived from the sun, are estimated to be neatly seven times as great as on the earth, and the apparent magnitude of the sun to a spectator on Mercury would be seven times greater than to us. Hence the sun would present to an inhabitant of that planet, with eyes like ours, an object of insufferable brightness; and all objects on the surface would be arrayed in a light more glorious than we can well imagine. (See Fig. 53.) The average heat on the greater portion of this planet would exceed that of boiling water, and therefore be incompatible with the existence both of an animal and a vegetable kingdom constituted like ours.

The motion of Mercury, in his revolution round the sun, is swifter than that of any other planet, being more than one hundred thousand miles every hour; whereas that of the earth is less than seventy thousand. Eighteen hundred miles every minute,—crossing the Atlantic ocean in less than two minutes,—this is a velocity of which we can form but a very inadequate conception, although, as we shall see hereafter, it is far less than comets sometimes exhibit.

Venus is regarded as the most beautiful of the planets, and is well known as the *morning and evening star*. The most ancient nations, indeed, did not recognise the morning and evening star as one and the same body, but supposed they were different planets, and accordingly gave them different names, calling the morning star Lucifer, and the evening star Hesperus. At her period of greatest splendor, Venus casts a shadow, and is sometimes visible in broad daylight. Her light is then estimated as equal to that of twenty stars of the first magnitude. In the equatorial regions of the earth, where the twilight is short, and Venus, at her greatest elongation, appears very high above the horizon, her splendors are said to be far more conspicuous than in our latitude.

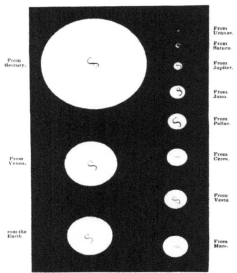

Fig. 53. APPARENT MAGNITUDES OF THE SUN, AS SEEN FROM THE DIFFERENT PLANETS.

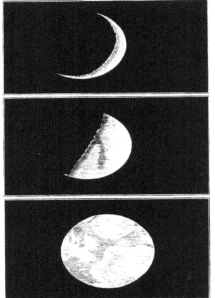

Figures 54, 55, 56. VENUS AND MARS.

Every eight years, Venus forms her conjunction with the sun in the same part of the heavens. Whatever appearances, therefore, arise from her position with respect to the earth and the sun, they are repeated every eight years, in nearly the same form.

Thus, every eight years, Venus is remarkably conspicuous, so as to be visible in the day-time, being then most favorably situated, on several accounts; namely, being nearest the earth, and at the point in her orbit where she gives her greatest brilliancy, that is, a little within the place of greatest elongation. This is the period for obtaining fine telescopic views of Venus, when she is seen with spots on her disk. Thus two figures of the annexed diagram (Fig. 54) represent Venus as seen near her inferior conjunction, and at the period of maximum brilliancy. The former situation is favorable for viewing her inequalities of surface, as indicated by the roughness of the line which separates the enlightened from the unenlightened part, (the *terminator.*) According to Schroeter, a German astronomer, Venus has mountains twenty-two miles high. Her mountains, however, are much more difficult to be seen than those of the moon.

The sun would appear, as seen from Venus, twice as large as on the earth, and its light and heat would be augmented in the same proportion. (See Fig. 53.) In many respects, however, the phenomena of this planet are similar to those of our own; and the general likeness between Venus and the earth, in regard to dimensions, revolutions, and seasons, is greater than exists between any other two bodies of the system.

I will only add to the present Letter a few words on the *transits* of the inferior planets.

The transit of Mercury or Venus is its passage across the sun's disk, as the moon passes over it in a solar eclipse. The planet is seen projected on the sun's disk in a small, black, round spot, moving slowly over the face of the sun. As the transit takes place only when the planet is in inferior conjunction, at which time her motion is retrograde, it is always from left to right; and, on account of its motion being retarded by the motion of the earth, (as was explained by Fig. 51, page 232,) it remains sometimes a long time on the solar disk. Mercury, when it makes its transit across the sun's centre, may remain on the sun from five to seven hours.

You may ask, why we do not observe this appearance every time one of the inferior planets comes into inferior conjunction, for then, of course, it passes between us and the sun. It must, indeed, at this time, cross the meridian at the same time with the sun; but, because its orbit is inclined to that of the sun, it may cross it (and generally does) a little above or a little below the sun. It is only when the conjunction takes place at or very near the point where the two orbits cross one another, that is, near the *node*, that a transit can occur. Thus, if the orbit of Mercury, N M R, Fig. 50, (page 231,) were in the same plane with the earth's orbit, (and of course with the sun's apparent orbit,) then, when the planet was at Q, in its inferior conjunction, the earth being at E, it would always be projected on the sun's disk at Z, on the concave sphere of the heavens, and a transit would happen at every inferior conjunction. But now let us take hold of the point R, and lift the circle which represents the orbit of Mercury upwards seven degrees, letting it turn upon the diameter *d b*; then, we may easily see that a spectator at E would project the planet higher in the heavens than the sun; and such would always be the case, except when the conjunction takes place at the node. Then the point of intersection of the two orbits being in one and the same plane, both bodies would be referred to the same point on the celestial sphere. As the sun, in his apparent revolution around the earth every year, passes through every point in the ecliptic, of course he must every year be at each of the points where the orbit of Mercury or Venus crosses the ecliptic, that is, at each of the nodes of one of these planets;[12] and as these nodes are on opposite sides of the ecliptic, consequently, the sun will pass through them at opposite seasons of the year, as in January and July, February and August. Now, should Mercury or Venus happen to come between us and the sun, just as the sun is passing one of the planet's nodes, a transit would happen. Hence the transits of Mercury take place in May and November, and those of Venus, in June and December.

Transits of Mercury occur more frequently than those of Venus. The periodic times of Mercury and the earth are so adjusted to each other, that Mercury performs nearly twenty-nine revolutions while the earth performs seven. If, therefore, the two bodies meet at the node in any given year, seven years afterwards they will meet nearly at the same node, and a transit may take place, accordingly, at intervals of seven years. But fifty-four revolutions of Mercury correspond still nearer to thirteen revolutions of the earth; and therefore a transit is still more probable after intervals of thirteen years. At intervals of thirty-three years, transits of Mercury are exceedingly probable, because in that time Mercury makes almost exactly one hundred and thirty-seven revolutions. Intermediate transits, however, may occur at the other node. Thus, transits of Mercury happened at the ascending node in 1815, and 1822, at intervals of seven years; and at the descending node in 1832, which will return in 1845, after thirteen years.

Transits of Venus are events of very unfrequent occurrence. Eight revolutions of the earth are completed in nearly the same time as thirteen revolutions of Venus; and hence two transits of Venus may occur after an interval of eight years, as was the case at the last return of the phenomenon, one transit having occurred in 1761, and another in 1769. But if a transit does not happen after eight years, it will not happen at the same node, until an interval of two hundred and thirty-five years: but intermediate transits may occur at the other node. The next transit of Venus will take place in

1874, being two hundred and thirty-five years after the first that was ever *observed*, which occurred in 1639. This was seen, for the first time by mortal eyes, by two youthful English astronomers, Horrox and Crabtree. Horrox was a young man of extraordinary promise, and indicated early talents for practical astronomy, which augured the highest eminence; but he died in the twenty-third year of his age. He was only twenty when the transit appeared, and he had made the calculations and observations, by which he was enabled to anticipate its arrival several years before. At the approach of the desired time for observing the transit, he received the sun's image through a telescope in a dark room upon a white piece of paper, and after waiting many hours with great impatience, (as his calculation did not lead him to a knowledge of the precise time of the occurrence,) at last, on the twenty-fourth of November, 1639, old style, at three and a quarter hours past twelve, just as he returned from church, he had the pleasure to find a large round spot near the limb of the sun's image. It moved slowly across the sun's disk, but had not entirely left it when the sun set.

The great interest attached by astronomers to a transit of Venus arises from its furnishing the most accurate means in our power of determining the *sun's horizontal parallax*,—an element of great importance, since it leads us to a knowledge of the distance of the earth from the sun, which again affords the means of estimating the distances of all the other planets, and possibly, of the fixed stars. Hence, in 1769, great efforts were made throughout the civilized world, under the patronage of different governments, to observe this phenomenon under circumstances the most favorable for determining the parallax of the sun.

The common methods of finding the parallax of a heavenly body cannot be relied on to a greater degree of accuracy than four seconds. In the case of the moon, whose greatest parallax amounts to about one degree, this deviation from absolute accuracy is not very material; but it amounts to nearly half the entire parallax of the sun.

If the sun and Venus were equally distant from us, they would be equally affected by parallax, as viewed by spectators in different parts of the earth, and hence their *relative* situation would not be altered by it; but since Venus, at the inferior conjunction, is only about one third as far off as the sun, her parallax is proportionally greater, and therefore spectators at distant points will see Venus projected on different parts of the solar disk, as the planet traverses the disk. Astronomers avail themselves of this circumstance to ascertain the sun's horizontal parallax, which they are enabled to do by comparing it with that of Venus, in a manner which, without a knowledge of trigonometry, you will not fully understand. In order to make the difference in the apparent places of Venus on the sun's disk as great as possible, very distant places are selected for observation. Thus, in the transits of 1761 and 1769, several of the European governments fitted out expensive expeditions to parts of the earth remote from each other. For this purpose, the celebrated Captain Cook, in 1769, went to the South Pacific Ocean, and observed the transit at the island of Otaheite, while others went to Lapland, for the same purpose, and others still, to many other parts of the globe. Thus, suppose two observers took their stations on opposite sides of the earth, as at A, and B, Fig. 57, page 242; at A, the planet V would be seen on the sun's disk at *a*, while at B, it would be seen at *b*.

The appearance of Venus on the sun's disk being that of a well-defined black spot, and the exactness with which the moment of external or internal contact may be determined, are circumstances favorable to the exactness of the result; and astronomers repose so much confidence in the estimation of the sun's horizontal parallax, as derived from observations on the transit of 1769, that this important element is thought to be ascertained within one tenth of a second. The general result of all these observations gives the sun's horizontal parallax eight seconds and six tenths,—a result which shows at once that the sun must be a great way off, since the semidiameter of the earth, a line nearly four thousand miles in length, would appear at the sun under an angle less than one four hundredth of a degree. During the transits of Venus over the sun's disk, in 1761 and 1769, a sort of penumbral light was observed around the planet, by several astronomers, which was thought to indicate an *atmosphere*. This appearance was particularly observable while the planet was coming on or going off the solar disk. The total immersion and emersion were not instantaneous; but as two drops of water, when about to separate, form a ligament between them, so there was a dark shade stretched out between Venus and the sun; and when the ligament broke, the planet seemed to have got about an eighth part of her diameter from the limb of the sun. The various accounts of the two transits abound with remarks like these, which indicate the existence of an atmosphere about Venus of nearly the density and extent of the earth's atmosphere. Similar proofs of the existence of an atmosphere around this planet are derived from appearances of twilight.

Fig. 57.

The elder astronomers imagined that they had discovered a *satellite* accompanying Venus in her transit. If Venus had in reality any satellite, the fact would be obvious at her transits, as, in some of them at least, it is probable that the satellite would be projected near the primary on the sun's disk; but later astronomers have searched in vain for any appearances of the kind, and the inference is, that former astronomers were deceived by some optical illusion.

LETTER XXI.

SUPERIOR PLANETS: MARS, JUPITER, SATURN, AND URANUS.

"With what an awful, world-revolving power,
Were first the unwieldy planets launched along
The illimitable void! There to remain
Amidst the flux of many thousand years,
That oft has swept the toiling race of men,
And all their labored monuments, away."—*Thomson.*

MERCURY AND VENUS, as we have seen, are always observed near the sun, and from this circumstance, as well as from the changes of magnitude and form which they undergo, we know that they have their orbits within that of the earth, and hence we call them *inferior* planets. On the other hand, Mars, Jupiter, Saturn, and Uranus, exhibit such appearances, at different times, as show that they revolve around the sun at a greater distance than the earth, and hence we denominate them *superior* planets. We know that they never come between us and the sun, because they never undergo those changes which Mercury and Venus, as well as the moon, sustain, in consequence of their coming into such a position. They, however, wander to the greatest angular distance from the sun, being sometimes seen one hundred and eighty degrees from him, so as to rise when the sun sets. All these different appearances must naturally result from their orbits' being exterior to that of the earth, as will be evident from the following representation. Let E, Fig. 58, page 244, be the earth, and M, one of the superior planets, Mars, for example, each body being seen in its path around the sun. At M, the planet would be in opposition to the sun, like the moon at the full; at Q and Q′, it would be seen ninety degrees off, or in quadrature; and at M′, in conjunction. We know, however, that this must be a superior and not an inferior conjunction, for the illuminated disk is still turned towards us; whereas, if it came between us and the sun, like Mercury, or Venus, in its inferior conjunction, its dark side would be presented to us.

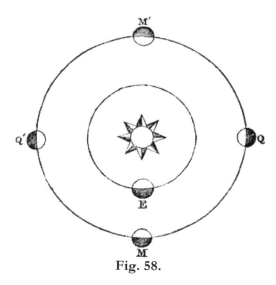

Fig. 58.

The superior planets do not exhibit to the telescope different phases, but, with a single exception, they always present the side that is turned towards the earth fully enlightened. This is owing to their great distance from the earth; for were the spectator to stand upon the sun, he would of course always have the illuminated side of each of the planets turned towards him; but so distant are all the superior planets, except Mars, that they are viewed by us very nearly, in the same manner as they would be if we actually stood on the sun. Mars, however, is sufficiently near to appear somewhat gibbous when at or near one of its quadratures. Thus, when the planet is at Q, it is plain that, of the hemisphere that is turned towards the earth, a small part is unilluminated.

Mars is a small planet, his diameter being only about half that of the earth, or four thousand two hundred miles. He also, at times, comes nearer to the earth than any other planet, except Venus. His *mean* distance from the sun is one hundred and forty-two millions of miles; but his orbit is so elliptical, that his distance varies much in different parts of his revolution. Mars is always very near the ecliptic, never varying from it more than two degrees. He is distinguished from all the planets by his deep red color, and fiery aspect; but his brightness and apparent magnitude vary much, at different times, being sometimes nearer to us than at others by the whole diameter of the earth's orbit; that is, by about one hundred and ninety millions of miles. When Mars is on the same side of the sun with the earth, or at his opposition, he comes within forty-seven millions of miles of the earth, and, rising about the time the sun sets, surprises us by his magnitude and splendor; but when he passes to the other side of the sun, to his superior conjunction, he dwindles to the appearance of a small star, being then two hundred and thirty-seven millions of miles from us. Thus, let M, Fig, 58, represent Mars in opposition, and M′, in the superior conjunction, while E represents the earth. It is obvious that, in the former situation, the planet must be nearer to the earth than in the latter, by the whole diameter of the earth's orbit. When viewed with a powerful telescope, the surface of Mars appears diversified with numerous varieties of light and shade. The region around the poles is marked by white spots, (see Fig. 56, page 237,) which vary their appearances with the changes of seasons in the planet. Hence Dr. Herschel conjectured that they were owing to ice and snow, which alternately accumulate and melt away, according as it is Winter or Summer, in that region. They are greatest and most conspicuous when that part of the planet has just emerged from a long Winter, and they gradually waste away, as they are exposed to the solar heat. Fig. 56, represents the planet, as exhibited, under the most favorable circumstances, to a powerful telescope, at the time when its gibbous form is strikingly obvious. It has been common to ascribe the ruddy light of Mars to an extensive and dense atmosphere, which was said to be distinctly indicated by the gradual diminution of light observed in a star, as it approaches very near to the planet, in undergoing an occultation; but more recent observations afford no such evidence of an atmosphere.

By observations on the spots, we learn that Mars revolves on his axis in very nearly the same time with the earth, (twenty-four hours thirty-nine minutes twenty-one seconds and three tenths,) and that the inclination of his axis to that of his orbit is also nearly the same, being thirty degrees eighteen minutes ten seconds and eight tenths. Hence the changes of day and night must be nearly the same there as here, and the seasons also very similar to ours. Since, however, the distance of Mars from the sun is one hundred and forty-two while that of the earth is only ninety-five millions of miles, the sun will appear more than twice as small on that planet as on ours, (see Fig. 53, page 236,) and its light and heat will be diminished in the same proportion. Only the equatorial regions, therefore, will be suitable for the existence of animals and vegetables.

Figures 59, 60. JUPITER AND SATURN.

The earth will be seen from Mars as an inferior planet, always near the sun, presenting appearances similar, in many respects, to those which Venus presents to us. It will be to that planet the evening and morning star, sung by their poets (if poets they have) with a like enthusiasm. The moon will attend the earth as a little star, being never seen further from her side than about the diameter under which we view the moon. To the telescope, the earth will exhibit phases similar to those of Venus; and, finally, she will, at long intervals, make her transits over the solar disk. Mean-while, Venus will stand to Mars in a relation similar to that of Mercury to us, revealing herself only when at the periods of her greatest elongation, and at all other times hiding herself within the solar blaze. Mercury will never be visible to an inhabitant of Mars.

Jupiter is distinguished from all the other planets by his great *magnitude.* His diameter is eighty-nine thousand miles, and his volume one thousand two hundred and eighty times that of the earth. His figure is strikingly spheroidal, the equatorial being more than six thousand miles longer than the polar diameter. Such a figure might naturally be expected from the rapidity of his diurnal rotation, which is accomplished in about ten hours. A place on the equator of Jupiter must turn twenty-seven times as fast as on the terrestrial equator. The distance of Jupiter from the sun is nearly four hundred and ninety millions of miles, and his revolution around the sun occupies nearly twelve years. Every thing appertaining to Jupiter is on a grand scale. A world in itself, equal in dimensions to twelve hundred and eighty of ours; the whole firmament rolling round it in the short space of ten hours, a movement so rapid that the eye could probably perceive the heavenly bodies to change their places every moment; its year dragging out a length of more than four thousand days, and more than ten thousand of its own days, while its nocturnal skies are lighted up with four brilliant moons;—these are some of the peculiarities which characterize this magnificent planet.

The view of Jupiter through a good telescope is one of the most splendid and interesting spectacles in astronomy. The disk expands into a large and bright orb, like the full moon; the spheroidal figure which theory assigns to revolving spheres, especially to those which turn with great velocity, is here palpably exhibited to the eye; across the disk, arranged in parallel stripes, are discerned several dusky bands, called *belts*; and four bright satellites, always in attendance, and ever varying their positions, compose a splendid retinue. Indeed, astronomers gaze with peculiar interest on Jupiter and his moons, as affording a miniature representation of the whole solar system, repeating, on a smaller scale, the same revolutions, and exemplifying more within the compass of our observation, the same laws as regulate the entire assemblage of sun and planets. Figure 59, facing page 247, gives a correct view of Jupiter, as exhibited to a powerful telescope in a clear evening. You will remark his flattened or spheroidal figure, the belts which appear in parallel stripes across his disk, and the four satellites, that are seen like little stars in a straight line with the equator of the planet.

The *belts of Jupiter* are variable in their number and dimensions. With the smaller telescopes only one or two are seen, and those across the equatorial regions; but with more powerful instruments, the number is increased, covering a large part of the entire disk. Different opinions have been entertained by astronomers respecting the cause of these belts; but they have generally been regarded as clouds formed in the atmosphere of the planet, agitated by winds, as is indicated by their frequent changes, and made to assume the form of belts parallel to the equator, like currents that circulate around our globe. Sir John Herschel supposes that the belts are not ranges of clouds, but portions of the planet itself, brought into view by the removal of clouds and mists, that exist in the atmosphere of the planet, through which are openings made by currents circulating around Jupiter.

The *satellites of Jupiter* may be seen with a telescope of very moderate powers. Even a common spyglass will enable us to discern them. Indeed, one or two of them have been occasionally seen with the naked eye. In the largest telescopes they severally appear as bright as Sirius. With such an instrument, the view of Jupiter, with his moons and belts, is truly a magnificent spectacle. As the orbits of the satellites do not deviate far from the plane of the ecliptic, and but little from the equator of the planet, they are usually seen in nearly a straight line with each other, extending across the central part of the disk. (See Fig. 59, facing page 247.)

Jupiter and his satellites exhibit in miniature all the phenomena of the solar system. The satellites perform, around their primary, revolutions very analogous to those which the planets perform around the sun, having, in like manner, motions alternately direct, stationary, and retrograde. They are all, with one exception, a little larger than the moon; and the second satellite, which is the smallest, is nearly as large as the moon, being two thousand and sixty-eight miles in diameter. They are all very small compared with the primary, the largest being only one twenty-sixth part of the primary. The outermost satellite extends to the distance from the planet of fourteen times his diameter. The whole system, therefore, occupies a region of space more than one million miles in breadth. Rapidity of motion, as well as greatness of dimensions, is characteristic of the system of Jupiter. I have already mentioned that the planet itself has a motion on its own axis much swifter than that of the earth, and the motions of the satellites are also much more rapid than that of the moon. The innermost, which is a little further off than the moon is from the earth, goes round its primary in about a day and three quarters; and the outermost occupies less than seventeen days.

The orbits of the satellites are nearly or quite circular, and deviate but little from the plane of the planet's equator, and of course are but slightly inclined to the plane of his orbit. They are therefore in a similar situation with respect to Jupiter, as the moon would be with respect to the earth, if her orbit nearly coincided with the ecliptic, in which case, she would undergo an eclipse at every opposition. The eclipses of Jupiter's satellites, in their general circumstances, are perfectly analogous to those of the moon, but in their details they differ in several particulars. Owing to the much greater distance of Jupiter from the sun, and its greater magnitude, the cone of its shadow is much longer and larger than that of the earth. On this account, as well as on account of the little inclination of their orbit to that of the primary, the three inner satellites of Jupiter pass through his shadow, and are totally eclipsed, at every revolution. The fourth satellite, owing to the greater inclination of its orbit, sometimes, though rarely, escapes eclipse, and sometimes merely grazes the limits of the shadow, or suffers a partial eclipse. These eclipses, moreover, are not seen, as is the case with those of the moon, from the centre of their motion, but from a remote station, and one whose situation with respect to the line of the shadow is variable. This makes no difference in the *times* of the eclipses, but it makes a very great one in their visibility, and in their apparent situations with respect to the planet at the moment of their entering or quitting the shadow.

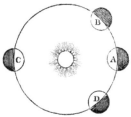

Fig. 61.

The eclipses of Jupiter's satellites present some curious phenomena, which you will easily understand by studying the following diagram. Let A, B, C, D, Fig. 61, represent the earth in different parts of its orbit; J, Jupiter, in his orbit, surrounded by his four satellites, the orbits of which are marked 1, 2, 3, 4. At *a*, the first satellite enters the shadow of the planet, emerges from it at *b*, and advances to its greatest elongation at *c*. The other satellites traverse the shadow in a similar manner. The apparent place, with respect to the planet, at which these eclipses will be seen to occur, will be altered by the position the earth happens at that moment to have in its orbit; but their appearances for any given night, as exhibited at Greenwich, are calculated and accurately laid down in the Nautical Almanac.

When one of the satellites is passing between Jupiter and the sun, it casts its shadow on the primary, as the moon casts its shadow on the earth in a solar eclipse. We see with the telescope the shadow traversing the disk. Sometimes, the satellite itself is seen projected on the disk; but, being illuminated as well as the primary, it is not so easily distinguished as Venus or Mercury, when seen on the sun's disk in one of their transits, since these bodies have their dark sides turned towards us; but the satellite is illuminated by the sun, as well as the primary, and therefore is not easily distinguishable from it.

The eclipses of Jupiter's satellites have been studied with great attention by astronomers, on account of their affording one of the easiest methods of determining the *longitude*. On this subject, Sir John Herschel remarks: "The discovery of Jupiter's satellites by Galileo, which was one of the first fruits of the invention of the telescope, forms one of the most memorable epochs in the history of astronomy. The first astronomical solution of the problem of 'the longitude,'—the most important problem for the interests of mankind that has ever been brought under the dominion of strict scientific principles,—dates immediately from this discovery. The final and conclusive establishment of the Copernican system of astronomy may also be considered as referable to the discovery and study of this exquisite miniature system, in which the laws of the planetary motions, as ascertained by Kepler, and especially that which connects their periods and distances, were speedily traced, and found to be satisfactorily maintained."

The entrance of one of Jupiter's satellites into the shadow of the primary, being seen like the entrance of the moon into the earth's shadow at the same moment of absolute time, at all places where the planet is visible, and being wholly independent of parallax, that is, presenting the same phenomenon to places remote from each other; being, moreover, predicted beforehand, with great accuracy, for the instant of its occurrence at Greenwich, and given in the Nautical Almanac; this would seem to be one of those events which are peculiarly adapted for finding the longitude. For you will recollect, that "any instantaneous appearance in the heavens, visible at the same moment of absolute time at any two places, may be employed for determining the difference of longitude between those places; for the difference in their local times, as indicated by clocks or chronometers, allowing fifteen degrees for every hour, will show their difference of longitude."

With respect to the method by the eclipses of Jupiter's satellites, it must be remarked, that the extinction of light in the satellite, at its immersion, and the recovery of its light at its emersion, are not instantaneous, but gradual; for the satellite, like the moon, occupies some time in entering into the shadow, or in emerging from it, which occasions a progressive diminution or increase of light. Two observers in the same room, observing with different telescopes the same eclipse, will frequently disagree, in noting its time, to the amount of fifteen or twenty seconds. Better methods, therefore, of finding the longitude, are now employed, although the facility with which the necessary observations can be made, and the little calculation required, still render this method eligible in many cases where extreme accuracy is not important. As a telescope is essential for observing an eclipse of one of the satellites, it is obvious that this method cannot be practised at sea, since the telescope cannot be used on board of ship, for want of the requisite steadiness.

The grand discovery of the *progressive motion of light* was first made by observations on the eclipses of Jupiter's satellites. In the year 1675, it was remarked by Roemer, a Danish astronomer, on comparing together observations of these eclipses during many successive years, that they take place sooner by about sixteen minutes, when the earth is on the same side of the sun with the planet, than when she is on the opposite side. The difference he ascribes to the progressive motion of light, which takes that time to pass through the diameter of the earth's orbit, making the velocity of light about one hundred and ninety-two thousand miles per second. So great a velocity startled astronomers at first, and produced some degree of distrust of this explanation of the phenomenon; but the subsequent discovery of what is called the aberration of light, led to an independent estimation of the velocity of light, with almost precisely the same result.

Few greater feats have ever been performed by the human mind, than to measure the speed of light,—a speed so great, as would carry it across the Atlantic Ocean in the sixty-fourth part of a second, and around the globe in less than the seventh part of a second! Thus has man applied his scale to the motions of an element, that literally leaps from world to world in the twinkling of an eye. This is one example of the great power which the invention of the telescope conferred on man.

Could we plant ourselves on the surface of this vast planet, we should see the same starry firmament expanding over our heads as we see now; and the same would be true if we could fly from one planetary world to another, until we made the circuit of them all; but the sun and the planetary system would present themselves to us under new and strange aspects. The sun himself would dwindle to one twenty-seventh of his present surface, (Fig. 53, facing page 236,) and afford a degree of light and heat proportionally diminished; Mercury, Venus, and even the Earth, would all disappear, being too near the sun to be visible; Mars would be as seldom seen as Mercury is by us, and constitute the only inferior planet. On the other hand, Saturn would shine with greatly augmented size and splendor. When in opposition to the sun, (at which time it comes nearest to Jupiter,) it would be a grand object, appearing larger than either Venus or Jupiter does to us. When, however, passing to the other side of the sun, through its superior conjunction, it would gradually diminish in size and brightness, and at length become much less than it ever appears to us, since it would then be four hundred millions of miles further from Jupiter than it ever is from us.

Although Jupiter comes four hundred millions of miles nearer to Uranus than the earth does, yet it is still thirteen hundred millions of miles distant from that planet. Hence the augmentation of the magnitude and light of Uranus would be barely sufficient to render it distinguishable by the naked eye. It appears, therefore, that Saturn is the peculiar ornament of the firmament of Jupiter, and would present to the telescope most interesting and sublime phenomena. As we owe the revelation of the system of Jupiter and his attendant worlds wholly to the telescope, and as the discovery and observation of them constituted a large portion of the glory of Galileo, I am now forcibly reminded of his labors, and will recur to his history, and finish the sketch which I commenced in a previous Letter.

LETTER XXII.

COPERNICUS.—GALILEO.

"They leave at length the nether gloom, and stand Before the portals of a better land; To happier plains they come, and fairer groves, The seats of those whom Heaven, benignant, loves; A brighter day, a bluer ether, spreads Its lucid depths above their favored heads; And, purged from mists that veil our earthly skies, Shine suns and stars unseen by mortal eyes."—*Virgil.*

IN order to appreciate the value of the contributions which Galileo made to astronomy, soon after the invention of the telescope, it is necessary to glance at the state of the science when he commenced his discoveries For many centuries, during the middle ages, a dark night had hung over astronomy, through which hardly a ray of light penetrated, when, in the eastern part of civilized Europe, a luminary appeared, that proved the harbinger of a bright and glorious day. This was Copernicus, a native of Thorn, in Prussia. He was born in 1473. Though destined for the profession of medicine, from his earliest years he displayed a great fondness and genius for mathematical studies, and pursued them with distinguished success in the University of Cracow. At the age of twenty-five years, he resorted to Italy, for the purpose of studying astronomy, where he resided a number of years. Thus prepared, he returned to his native country, and, having acquired an ecclesiastical living that was adequate to his support in his frugal mode of life, he established himself at Frauenberg, a small town near the mouth of the Vistula, where he spent nearly forty years in observing the heavens, and meditating on the celestial motions. He occupied the upper part of a humble farm-house, through the roof of which he could find access to an unobstructed sky, and there he carried on his observations. His instruments, however, were few and imperfect, and it does not appear that he added any thing to the art of practical astronomy. This was reserved for Tycho Brahe, who came a half a century after him. Nor did Copernicus enrich the science with any important discoveries. It was not so much his genius or taste to search for new bodies, or new phenomena among the stars, as it was to explain the reasons of the most obvious and well-known appearances and motions of the heavenly bodies. With this view, he gave his mind to long-continued and profound meditation.

Copernicus tells us that he was first led to think that the apparent motions of the heavenly bodies, in their diurnal revolution, were owing to the real motion of the earth in the opposite direction, from observing instances of the same kind among terrestrial objects; as when the shore seems to the mariner to recede, as he rapidly sails from it; and as trees and other objects seem to glide by us, when, on riding swiftly past them, we lose the consciousness of our own motion. He was also smitten with the *simplicity* prevalent in all the works and operations of Nature, which is more and more conspicuous the more they are understood; and he hence concluded that the planets do not move in the complicated paths which most preceding astronomers assigned to them. I shall explain to you, hereafter, the details of his system. I need only at present remind you that the hypothesis which he espoused and defended, (being substantially the same as that proposed by Pythagoras, five hundred years before the Christian era,) supposes, first, that the apparent movements of the sun by day, and of the moon and stars by night, from east to west, result from the actual revolution of the earth on its own axis from west to east; and, secondly, that the earth and all the planets revolve about the sun in circular orbits. This hypothesis, when he first assumed it, was with him, as it had been with Pythagoras, little more than mere conjecture. The arguments by which its truth was to be finally established were not yet developed, and could not be, without the aid of the telescope, which was not yet invented. Upon this hypothesis, however, he set out to explain all the phenomena of the visible heavens,—as the diurnal revolutions of the sun, moon, and stars, the slow progress of the planets through the signs of the zodiac, and the numerous irregularities to which the planetary motions are subject. These last are apparently so capricious,—being for some time forward, then stationary, then backward, then stationary again, and finally direct, a second time, in the order of the signs, and constantly varying in the velocity of their movements,—that nothing but long-continued and severe meditation could have solved all these appearances, in conformity with the idea that each planet is pursuing its simple way all the while in a circle around the sun. Although, therefore, Pythagoras fathomed the profound doctrine that the sun is the centre around which the earth and all the planets revolve, yet we have no evidence that he ever solved the irregular motions of the planets in conformity with his hypothesis, although the explanation of the diurnal revolution of the heavens, by that hypothesis, involved no difficulty. Ignorant as Copernicus was of the principle of gravitation, and of most of the laws of motion, he could go but little way in following out the consequences of his own hypothesis; and all that can be claimed for him is, that he solved, by means of it, most of the common phenomena of the celestial motions. He indeed got upon the road to truth, and advanced some way in its sure path; but he was able to adduce but few independent proofs, to show that it was the truth. It was only quite at the close of his life that he published his system to the world, and that only at the urgent request of his friends; anticipating, perhaps, the opposition of a bigoted priesthood, whose fury was afterwards poured upon the head of Galileo, for maintaining the same doctrines.

Although, therefore, the system of Copernicus afforded an explanation of the celestial motions, far more simple and rational than the previous systems which made the earth the centre of those motions, yet this fact alone was not sufficient to compel the assent of astronomers; for the greater part, to say the least, of the same phenomena, could be explained on either hypothesis. With the old doctrine astronomers were already familiar, a circumstance which made it seem easier; while the new doctrines would seem more difficult, from their being imperfectly understood. Accordingly, for nearly a century after the publication of the system of Copernicus, he gained few disciples. Tycho Brahe rejected it, and proposed one of his own, of which I shall hereafter give you some account; and it would probably have fallen quite into oblivion, had not the observations of Galileo, with his newly-invented telescope, brought to light innumerable proofs of its truth, far more cogent than any which Copernicus himself had been able to devise.

Galileo no sooner had completed his telescope, and directed it to the heavens, than a world of wonders suddenly burst upon his enraptured sight. Pointing it to the moon, he was presented with a sight of her mottled disk, and of her mountains and valleys. The sun exhibited his spots; Venus, her phases; and Jupiter, his expanded orb, and his retinue of moons. These last he named, in honor of his patron, Cosmo d'Medici, *Medicean stars*. So great was this honor deemed of associating one's name with the stars, that express application was made to Galileo, by the court of France, to award this distinction to the reigning Monarch, Henry the Fourth, with plain intimations, that by so doing he would render himself and his family rich and powerful for ever.

Galileo published the result of his discoveries in a paper, denominated 'Nuncius Sidereus,' the 'Messenger of the Stars.' In that ignorant and marvellous age, this publication produced a wonderful excitement. "Many doubted, many positively refused to believe, so novel an announcement; all were struck with the greatest astonishment, according to their respective opinions, either at the new view of the universe thus offered to them, or at the high audacity of Galileo, in inventing such fables." Even Kepler, the great German astronomer, of whom I shall tell you more by and by, wrote to Galileo, and desired him to supply him with arguments, by which he might answer the objections to these pretended discoveries with which he was continually assailed. Galileo answered him as follows: "In the first place, I return you my thanks that you first, and almost alone, before the question had been sifted, (such is your candor, and the loftiness of your mind,) put faith in my assertions. You tell me you have some telescopes, but not sufficiently good to magnify distant objects with clearness, and that you anxiously expect a sight of mine, which magnifies images more than a thousand times. It is mine no longer, for the Grand Duke of Tuscany has asked it of me, and intends to lay it up in his museum, among his most rare and precious curiosities, in eternal remembrance of the invention.

"You ask, my dear Kepler, for other testimonies. I produce, for one, the Grand Duke, who, after observing the Medicean planets several times with me at Pisa, during the last months, made me a present, at parting, of more than a thousand florins, and has now invited me to attach myself to him, with the annual salary of one thousand florins, and with the title of 'Philosopher and Principal Mathematician to His Highness;' without the duties of any office to perform, but with the most complete leisure. I produce, for another witness, myself, who, although already endowed in this College with the noble salary of one thousand florins, such as no professor of mathematics ever before received, and which I might securely enjoy during my life, even if these planets should deceive me and should disappear, yet quit this situation, and take me where want and disgrace will be my punishment, should I prove to have been mistaken."

The learned professors in the universities, who, in those days, were unaccustomed to employ their senses in inquiring into the phenomena of Nature, but satisfied themselves with the authority of Aristotle, on all subjects, were among the most incredulous with respect to the discoveries of Galileo. "Oh, my dear Kepler," says Galileo, "how I wish that we could have one hearty laugh together. Here, at Padua, is the principal Professor of Philosophy, whom I have repeatedly and urgently requested to look at the moon and planets through my glass, which he pertinaciously refuses to do. Why are you not here? What shouts of laughter we should have at this glorious folly, and to hear the Professor of Philosophy at Pisa laboring before the Grand Duke, with logical arguments, as if with magical incantations, to charm the new planets out of the sky."

The following argument by Sizzi, a contemporary astronomer of some note, to prove that there can be only seven planets, is a specimen of the logic with which Galileo was assailed. "There are seven windows given to animals in the domicile of the head, through which the air is admitted to the tabernacle of the body, to enlighten, to warm, and to nourish it; which windows are the principal parts of the microcosm, or little world,—two nostrils, two eyes, two ears, and one mouth. So in the heavens, as in a macrocosm, or great world, there are two favorable stars, Jupiter and Venus; two unpropitious, Mars and Saturn; two luminaries, the Sun and

Moon; and Mercury alone, undecided and indifferent. From which, and from many other phenomena of Nature, such as the seven metals, &c., which it were tedious to enumerate, we gather that the number of planets is necessarily seven. Moreover, the satellites are invisible to the naked eye, and therefore can exercise no influence over the earth, and therefore would be useless, and therefore do not exist. Besides, as well the Jews and other ancient nations, as modern Europeans, have adopted the division of the week into seven days, and have named them from the seven planets. Now, if we increase the number of planets, this whole system falls to the ground."

When, at length, the astronomers of the schools found it useless to deny the fact that Jupiter is attended by smaller bodies, which revolve around him, they shifted their ground of warfare, and asserted that Galileo had not told the whole truth; that there were not merely *four* satellites, but a still greater number; one said five; another, nine; and another, twelve; but, in a little time, Jupiter moved forward in his orbit, and left all behind him, save the four Medicean stars.

It had been objected to the Copernican system, that were Venus a body which revolved around the sun in an orbit interior to that of the earth, she would undergo changes similar to those of the moon. As no such changes could be detected by the naked eye, no satisfactory answer could be given to this objection; but the telescope set all right, by showing, in fact, the phases of Venus. The same instrument, disclosed, also, in the system of Jupiter and his moons, a miniature exhibition of the solar system itself. It showed the actual existence of the motion of a number of bodies around one central orb, exactly similar to that which was predicated of the sun and planets. Every one, therefore, of these new and interesting discoveries, helped to confirm the truth of the system of Copernicus.

But a fearful cloud was now rising over Galileo, which spread itself, and grew darker every hour. The Church of Rome had taken alarm at the new doctrines respecting the earth's motion, as contrary to the declarations of the Bible, and a formidable difficulty presented itself, namely, how to publish and defend these doctrines, without invoking the terrible punishments inflicted by the Inquisition on heretics. No work could be printed without license from the court of Rome; and any opinions supposed to be held and much more known to be taught by any one, which, by an ignorant and superstitious priesthood, could be interpreted as contrary to Scripture, would expose the offender to the severest punishments, even to imprisonment, scourging, and death. We, who live in an age so distinguished for freedom of thought and opinion, can form but a very inadequate conception of the bondage in which the minds of men were held by the chains of the Inquisition. It was necessary, therefore, for

Galileo to proceed with the greatest caution in promulgating truths which his own discoveries had confirmed. He did not, like the Christian martyrs, proclaim the truth in the face of persecutions and tortures; but while he sought to give currency to the Copernican doctrines, he labored, at the same time, by cunning artifices, to blind the ecclesiastics to his real designs, and thus to escape the effects of their hostility.

Before Galileo published his doctrines in form, he had expressed himself so freely, as to have excited much alarm among the ecclesiastics. One of them preached publicly against him, taking for his text, the passage, "Ye men of Galilee, why stand ye here gazing up into heaven?" He therefore thought it prudent to resort to Rome, and confront his enemies face to face. A contemporary describes his appearance there in the following terms, in a letter addressed to a Romish Cardinal: "Your Eminence would be delighted with Galileo, if you heard him holding forth, as he often does, in the midst of fifteen or twenty, all violently attacking him, sometimes in one house, sometimes in another. But he is armed after such fashion, that he laughs all of them to scorn; and even if the novelty of his opinions prevents entire persuasion, at least he convicts of emptiness most of the arguments with which his adversaries endeavor to overwhelm him."

In 1616, Galileo, as he himself states, had a most gracious audience of the Pope, Paul the Fifth, which lasted for nearly an hour, at the end of which his Holiness assured him, that the Congregation were no longer in a humor to listen lightly to calumnies against him, and that so long as he occupied the Papal chair, Galileo might think himself out of all danger. Nevertheless, he was not allowed to return home, without receiving formal notice not to teach the opinions of Copernicus, "that the sun is in the centre of the system, and that the earth moves about it," from that time forward, in any manner.

Galileo had a most sarcastic vein, and often rallied his persecutors with the keenest irony. This he exhibited, some time after quitting Rome, in an epistle which he sent to the Arch Duke Leopold, accompanying his 'Theory of the Tides.' "This theory," says he, "occurred to me when in Rome, whilst the theologians were debating on the prohibition of Copernicus's book, and of the opinion maintained in it of the motion of the earth, which I at that time believed; until it pleased those gentlemen to suspend the book, and to declare the opinion false and repugnant to the Holy Scriptures. Now, as I know how well it becomes me to obey and believe the decisions of my superiors, which proceed out of more profound knowledge than the weakness of my intellect can attain to, this theory, which I send you, which is founded on the motion of the earth, I now look upon as a fiction and a dream, and beg your Highness to receive it as such. But, as poets often learn to prize the creations of their fancy, so, in like manner, do I set some value on this absurdity of mine. It is true, that when I sketched this little work, I did hope that Copernicus would not, after eighty years, be convicted of error; and I had intended to develope and amplify it further; but a voice from heaven suddenly awakened me, and at once annihilated all my confused and entangled fancies."

It is difficult, however, sometimes to decide whether the language of Galileo is ironical, or whether he uses it with subtlety, with the hope of evading the anathemas of the Inquisition. Thus he ends one of his writings with the following passage: "In conclusion, since the motion attributed to the earth, which I, as a pious and Catholic person, consider most false, and not to exist, accommodates itself so well to explain so many and such different phenomena, I shall not feel sure that, false as it is, it may not just as deludingly correspond with the phenomena of comets."

In the year 1624, soon after the accession of Urban the Eighth to the Pontifical chair, Galileo went to Rome again, to offer his congratulations to the new Pope, as well as to propitiate his favor. He seems to have been received with unexpected cordiality; and, on his departure, the Pope commended him to the good graces of Ferdinand, Grand Duke of Tuscany, in the following terms: "We find in him not only literary distinction, but also the love of piety, and he is strong in those qualities by which Pontifical good-will is easily obtained. And now, when he has been brought to this city, to congratulate Us on Our elevation, We have lovingly embraced him; nor can We suffer him to return to the country whither your liberality recalls him, without an ample provision of Pontifical love. And that you may know how dear he is to Us, we have willed to give him this honorable testimonial of virtue and piety. And We further signify, that every benefit which you shall confer upon him will conduce to Our gratification."

In the year 1630, Galileo finished a great work, on which he had been long engaged, entitled, 'The Dialogue on the Ptolemaic and Copernican Systems.' From the notion which prevailed, that he still countenanced the Copernican doctrine of the earth's motion, which had been condemned as heretical, it was some time before he could obtain permission from the Inquisitors (whose license was necessary to every book) to publish it. This he was able to do, only by employing again that duplicity or artifice which would throw dust in the eyes of the vain and superstitious priesthood. In 1632, the work appeared under the following title: 'A Dialogue, by Galileo Galilei, Extraordinary Mathematician of the University of Pisa, and Principal Philosopher and Mathematician of the Most Serene Grand Duke of Tuscany; in which, in a Conversation of four days, are discussed the two principal Systems of the World, the Ptolemaic and Copernican, indeterminately proposing the Philosophical Arguments as well on one side as on the other.' The subtle disguise which he wore, may be seen from the following extract from his 'Introduction,' addressed 'To the discreet Reader.'

"Some years ago, a salutary edict was promulgated at Rome, which, in order to obviate the perilous scandals of the present age, enjoined an opportune silence on the Pythagorean opinion of the earth's motion. Some were not wanting, who rashly asserted that this decree originated, not in a judicious examination, but in ill-informed passion; and complaints were heard, that counsellors totally inexperienced in astronomical observations ought not, by hasty prohibitions, to clip the wings of speculative minds. My zeal could not keep silence when I heard these rash lamentations, and I thought it proper, as being fully informed with regard to that most prudent determination, to appear publicly on the theatre of the world, as a witness of the actual truth. I happened at that time to be in Rome: I was admitted to the audiences, and enjoyed the approbation, of the most eminent prelates of that court; nor did the publication of that decree occur without my receiving some prior intimation of it. Wherefore, it is my intention, in this present work, to show to foreign nations, that as much is known of this matter in Italy, and particularly in Rome, as ultramontane diligence can ever have formed any notion of, and collecting together all my own speculations on the Copernican system, to give them to understand that the knowledge of all these preceded the Roman censures; and that from this country proceed not only dogmas for the salvation of the soul, but also ingenious discoveries for the gratification of the understanding. With this object, I have taken up in the 'Dialogue' the Copernican side of the question, treating it as a pure mathematical hypothesis; and endeavoring, in every artificial manner, to represent it as having the advantage, not over the opinion of the stability of the earth absolutely, but according to the manner in which that opinion is defended by some, who indeed profess to be Aristotelians, but retain only the name, and are contented, without improvement, to worship shadows, not philosophizing with their own reason, but only from the recollection of the four principles imperfectly understood."

Although the Pope himself, as well as the Inquisitors, had examined Galileo's manuscript, and, not having the sagacity to detect the real motives of the author, had consented to its publication, yet, when the book was out, the enemies of Galileo found means to alarm the court of Rome, and Galileo was summoned to appear before the Inquisition. The philosopher was then seventy years old, and very infirm, and it was with great difficulty that he performed the journey. His unequalled dignity and celebrity, however, commanded the involuntary respect of the tribunal before which he was summoned, which they manifested by permitting him to reside at the palace of his friend, the Tuscan Ambassador; and when it became necessary, in the course of the inquiry, to examine him in person, although his removal to the Holy Office was then insisted upon, yet he was not, like other heretics, committed to close and solitary confinement. On the contrary, he was lodged in the apartments of the Head of the Inquisition, where he was allowed the attendance of his own servant, who was also permitted to sleep in an adjoining room, and to come and go at pleasure. These were deemed extraordinary indulgences in an age when the punishment of heretics usually began before their trial.

About four months after Galileo's arrival in Rome, he was summoned to the Holy Office. He was detained there during the whole of that day; and on the next day was conducted, in a penitential dress, to the Convent of Minerva, where the Cardinals and Prelates, his judges, were assembled for the purpose of passing judgement upon him, by which this venerable old man was solemnly called upon to renounce and abjure, as impious and heretical, the opinions which his whole existence had been consecrated to form and strengthen. Probably there is not a more curious document in the history of science, than that which contains the sentence of the Inquisition on Galileo, and his consequent abjuration. It teaches us so much, both of the darkness and bigotry of the terrible Inquisition, and of the sufferings encountered by those early martyrs of science, that I will transcribe for your perusal, from the excellent 'Life of Galileo' in the 'Library of Useful Knowledge,' (from which I have borrowed much already,) the entire record of this transaction. The sentence of the Inquisition is as follows:

"We, the undersigned, by the grace of God, Cardinals of the Holy Roman Church, Inquisitors General throughout the whole Christian Republic, Special Deputies of the Holy Apostolical Chair against heretical depravity:

"Whereas, you, Galileo, son of the late Vincenzo Galilei of Florence, aged seventy years, were denounced in 1615, to this Holy Office, for holding as true a false doctrine taught by many, namely, that the sun is immovable in the centre of the world, and that the earth moves, and also with a diurnal motion; also, for having pupils which you instructed in the same opinions; also, for maintaining a correspondence on the same with some German mathematicians; also, for publishing certain letters on the solar spots, in which you developed the same doctrine as true; also, for answering the objections which were continually produced from the Holy Scriptures, by glozing the said Scriptures, according to your own meaning; and whereas, thereupon was produced the copy of a writing, in form of a letter, professedly written by you to a person formerly your pupil, in which, following the hypothesis of Copernicus, you include several propositions contrary to the true sense and authority of the Holy Scriptures: therefore, this Holy Tribunal, being desirous of providing against the disorder and mischief which was thence proceeding and increasing, to the detriment of the holy faith, by the desire of His Holiness, and of the Most Eminent Lords Cardinals of this supreme and universal Inquisition, the two propositions of the stability of the sun, and motion of the earth, were *qualified* by the *Theological Qualifiers*, as follows:

"1. The proposition that the sun is in the centre of the world, and immovable from its place, is absurd, philosophically false, and formally heretical; because it is expressly contrary to the Holy Scriptures.

"2. The proposition that the earth is not the centre of the world, nor immovable, but that it moves, and also with a diurnal motion, is also absurd, philosophically false, and, theologically considered, equally erroneous in faith.

"But whereas, being pleased at that time to deal mildly with you, it was decreed in the Holy Congregation, held before His Holiness on the twenty-fifth day of February, 1616, that His Eminence the Lord Cardinal Bellarmine should enjoin you to give up altogether the said false doctrine; if you should refuse, that you should be ordered by the Commissary of the Holy Office to relinquish it, not to teach it to others, nor to defend it, and in default of the acquiescence, that you should be imprisoned; and in execution of this decree, on the following day, at the palace, in presence of His Eminence the said Lord Cardinal Bellarmine, after you had been mildly admonished by the said Lord Cardinal, you were commanded by the acting Commissary of the Holy Office, before a notary and witnesses, to relinquish altogether the said false opinion, and in future neither to defend nor teach it in any manner, neither verbally nor in writing, and upon your promising obedience, you were dismissed.

"And, in order that so pernicious a doctrine might be altogether rooted out, nor insinuate itself further to the heavy detriment of the Catholic truth, a decree emanated from the Holy Congregation of the Index, prohibiting the books which treat of this doctrine; and it was declared false, and altogether contrary to the Holy and Divine Scripture.

"And whereas, a book has since appeared, published at Florence last year, the title of which showed that you were the author, which title is, '*The Dialogue of Galileo Galilei, on the two principal Systems of the World, the Ptolemaic and Copernican;*' and whereas, the Holy Congregation has heard that, in consequence of printing the said book, the false opinion of the earth's motion and stability of the sun is daily gaining ground; the said book has been taken into careful consideration, and in it has been detected a glaring violation of the said order, which had been intimated to you; inasmuch as in this book you have defended the said opinion, already, and in your presence, condemned; although in the said book you labor, with many circumlocutions, to induce the belief that it is left by you undecided, and in express terms probable; which is equally a very grave error, since an opinion can in no way be probable which has been already declared and finally determined contrary to the Divine Scripture. Therefore, by Our order, you have been cited to this Holy Office, where, on your examination upon oath, you have acknowledged the said book as written and printed by you. You also confessed that you began to write the said book ten or twelve years ago, after the order aforesaid had been given. Also, that you demanded license to publish it, but without signifying to those who granted you this permission, that you had been commanded not to hold, defend, or teach, the said doctrine in any manner. You also confessed, that the style of said book was, in many places, so composed, that the reader might think the arguments adduced on the false side to be so worded, as more effectually to entangle the understanding than to be easily solved, alleging, in excuse, that you have thus run into an error, foreign (as you say) to your intention, from writing in the form of a dialogue, and in consequence of the natural complacency which every one feels with regard to his own subtilties, and in showing himself more skilful than the generality of mankind in contriving, even in favor of false propositions, ingenious and apparently probable arguments.

"And, upon a convenient time being given you for making your defence, you produced a certificate in the handwriting of His Eminence, the Lord Cardinal Bellarmine, procured, as you said, by yourself, that you might defend yourself against the calumnies of your enemies, who reported that you had abjured your opinions, and had been punished by the Holy Office; in which certificate it is declared, that you had not abjured, nor had been punished, but merely that the declaration made by his Holiness, and promulgated by the Holy Congregation of the Index, had been announced to you, which declares that the opinion of the motion of the earth, and stability of the sun, is contrary to the Holy Scriptures, and therefore cannot be held or defended. Wherefore, since no mention is there made of two articles of the order, to wit, the order 'not to teach,' and 'in any manner,' you argued that we ought to believe that, in the lapse of fourteen or sixteen years, they had escaped your memory, and that this was also the reason why you were silent as to the order, when you sought permission to publish your book, and that this is said by you, not to excuse your error, but that it may be attributed to vain-glorious ambition rather than to malice. But this very certificate, produced on your behalf, has greatly aggravated your offence, since it is therein declared, that the said opinion is contrary to the Holy Scriptures, and yet you have dared to treat of it, and to argue that it is probable; nor is there any extenuation in the license artfully and cunningly extorted by you, since you did not intimate the command imposed upon you. But whereas, it appeared to Us that you had not disclosed the whole truth with regard to your intentions, We thought it necessary to proceed to the rigorous examination of you, in which (without any prejudice to what you had confessed, and which is above detailed against you, with regard to your said intention) you answered like a good Catholic.

"Therefore, having seen and maturely considered the merits of your cause, with your said confessions and excuses, and every thing else which ought to be seen and considered, We have come to the underwritten final sentence against you:

"Invoking, therefore, the most holy name of our Lord Jesus Christ, and of his Most Glorious Virgin Mother, Mary, by this Our final sentence, which, sitting in council and judgement for the tribunal of the Reverend Masters of Sacred Theology, and Doctors of both Laws, Our Assessors, We put forth in this writing touching the matters and controversies before Us, between the Magnificent Charles Sincerus, Doctor of both Laws, Fiscal Proctor of this Holy Office, of the one part, and you, Galileo Galilei, an examined and confessed criminal from this present writing now in progress, as above, of the other part, We pronounce, judge, and declare, that you, the said Galileo, by reason of these things which have been detailed in the course of this writing, and which, as above, you have confessed, have rendered yourself vehemently suspected, by this Holy Office, of heresy; that is to say, that you believe and hold the false doctrine, and contrary to the Holy and Divine Scriptures, namely, that the sun is the centre of the world, and that it does not move from east to west, and that the earth does move, and is not the centre of the world; also, that an opinion can be held and supported, as probable, after it has been declared and finally decreed contrary to the Holy Scripture, and consequently, that you have incurred all the censures and penalties enjoined and promulgated in the sacred canons, and other general and particular constitutions against delinquents of this description. From which it is Our pleasure that you be absolved, provided that, with a sincere heart and unfeigned faith, in Our presence, you abjure, curse, and detest, the said errors and heresies, and every other error and heresy, contrary to the Catholic and Apostolic Church of Rome, in the form now shown to you.

"But that your grievous and pernicious error and transgression may not go altogether unpunished, and that you may be made more cautious in future, and may be a warning to others to abstain from delinquencies of this sort, We decree, that the book of the Dialogues of Galileo Galilei be prohibited by a public edict, and We condemn you to the formal prison of this Holy Office for a period determinable at Our pleasure; and, by way of salutary penance, We order you, during the next three years, to recite, once a week, the seven penitential psalms, reserving to Ourselves the power of moderating, commuting, or taking off the whole or part of the said punishment, or penance.

"And so We say, pronounce, and by Our sentence declare, decree, and reserve, in this and in every other better form and manner, which lawfully We may and can use. So We, the subscribing Cardinals, pronounce." [Subscribed by seven Cardinals.]

In conformity with the foregoing sentence, Galileo was made to kneel before the Inquisition, and make the following *Abjuration*:

"I, Galileo Galilei, son of the late Vincenzo Galilei, of Florence, aged seventy years, being brought personally to judgement, and kneeling before you, Most Eminent and Most Reverend Lords Cardinals, General Inquisitors of the Universal Christian Republic against heretical depravity, having before my eyes the Holy Gospels, which I touch with my own hands, swear, that I have always believed, and with the help of God will in future believe, every article which the Holy Catholic and Apostolic Church of Rome holds, teaches, and preaches. But because I had been enjoined, by this Holy Office, altogether to abandon the false opinion which maintains that the sun is the centre and immovable, and forbidden to hold, defend, or teach, the said false doctrine, in any manner: and after it had been signified to me that the said doctrine is repugnant to the Holy Scripture, I have written and printed a book, in which I treat of the same doctrine now condemned, and adduce reasons with great force in support of the same, without giving any solution, and therefore have been judged grievously suspected of heresy; that is to say, that I held and believed that the sun is the centre of the world and immovable, and that the earth is not the centre and movable; willing, therefore, to remove from the minds of Your Eminences, and of every Catholic Christian, this vehement suspicion rightfully entertained towards me, with a sincere heart and unfeigned faith, I abjure, curse, and defeat, the said errors and heresies, and generally every other error and sect contrary to the said Holy Church; and I swear, that I will never more in future say or assert any thing, verbally or in writing, which may give rise to a similar suspicion of me: but if I shall know any heretic, or any one suspected of heresy, that I will denounce him to this Holy Office, or to the Inquisitor and Ordinary of the place in which I may be. I swear, moreover, and promise, that I will fulfil and observe fully, all the penances which have been or shall be laid on me by this Holy Office. But if it shall happen that I violate any of my said promises, oaths, and protestations, (which God avert!) I subject myself to all the pains and punishments which have been decreed and promulgated by the sacred canons, and other general and particular constitutions, against delinquents of this description. So may God help me, and his Holy Gospels, which I touch with my own hands. I, the above-named Galileo Galilei, have abjured, sworn, promised, and bound myself, as above; and in witness thereof, with my own hand have subscribed this present writing of my abjuration, which I have recited, word for word.

"At Rome, in the Convent of Minerva, twenty-second June, 1633, I, Galileo Galilei, have abjured as above, with my own hand."

From the court Galileo was conducted to prison, to be immured for life in one of the dungeons of the Inquisition. His sentence was afterwards mitigated, and he was permitted to return to Florence; but the humiliation to which he had been subjected pressed heavily on his spirits, beset as he was with infirmities, and totally blind, and he never more talked or wrote on the subject of astronomy.

There was enough in the character of Galileo to command a high admiration. There was much, also, in his sufferings in the cause of science, to excite the deepest sympathy, and even compassion. He is moreover universally represented to have been a man of great equanimity, and of a noble and generous disposition. No scientific character of the age, or perhaps of any age, forms a structure of finer proportions, or wears in a higher degree the grace of symmetry. Still, we cannot approve of his employing artifice in the promulgation of truth; and we are compelled to lament that his lofty spirit bowed in the final conflict. How far, therefore, he sinks below the dignity of the Christian martyr! "At the age of seventy," says Dr. Brewster, in his life of Sir Isaac Newton, "on his bended knees, and with his right hand resting on the Holy Evangelists, did this patriarch of science avow his present and past belief in the dogmas of the Romish Church, abandon as false and heretical the doctrine of the earth's motion and of the sun's immobility, and pledge himself to denounce to the Inquisition any other person who was even suspected of heresy. He abjured, cursed, and detested, those eternal and immutable truths which the Almighty had permitted him to be the first to establish. Had Galileo but added the courage of the martyr to the wisdom of the sage; had he carried the glance of his indignant eye round the circle of his judges; had he lifted his hands to heaven, and called the living God to witness the truth and immutability of his opinions; the bigotry of his enemies would have been disarmed, and science would have enjoyed a memorable triumph."

LETTER XXIII.

SATURN.—URANUS.—ASTEROIDS.

"Into the Heaven of Heavens
I have presumed, An earthly guest, and drawn empyreal air."—*Milton.*

THE consideration of the system of Jupiter and his satellites led us to review the interesting history of Galileo, who first, by means of the telescope, disclosed the knowledge of that system to the world. I will now proceed with the other superior planets.

Saturn, as well as Jupiter, has within itself a system on a scale of great magnificence. In size it is next to Jupiter the largest of the planets, being seventy-nine thousand miles in diameter, or about one thousand times as large as the earth. It has likewise belts on its surface, and is attended by seven satellites. But a still more wonderful appendage is its *Ring*, a broad wheel, encompassing the planet at a great distance from it. As Saturn is nine hundred millions of miles from us, we require a more powerful telescope to see his glories, in all their magnificence, than we do to enjoy a full view of the system of Jupiter. When we are privileged with a view of Saturn, in his most favorable positions, through a telescope of the larger class, the mechanism appears more wonderful than even that of Jupiter.

Saturn's ring, when viewed with telescopes of a high power, is found to consist of two concentred rings, separated from each other by a dark space. Although this division of the rings appears to us, on account of our immense distance, as only a fine line, yet it is, in reality, an interval of not less than eighteen hundred miles. The dimensions of the whole system are, in round numbers, as follows:

	Miles.
Diameter of the planet,	79,000
From the surface of the planet to the inner ring,	20,000
Breadth of the inner ring,	17,000
Interval between the rings,	1,800
Breadth of the outer ring,	10,500
Extreme dimensions from outside to outside,	176,000

Figure 60, facing page 247, represents Saturn, as it appears to a powerful telescope, surrounded by its rings, and having its body striped with dark belts, somewhat similar, but broader and less strongly marked than those of Jupiter. In telescopes of inferior power, but still sufficient to see the ring distinctly, we should scarcely discern the belts at all. We might, however, observe the shadow cast upon the ring by the planet, (as seen in the figure on the right, on the upper side;) and, in favorable situations of the planet, we might discern glimpses of the shadow of the ring on the body of the planet, on the lower side beneath the ring. To see the division of the ring and the satellites requires a better telescope than is in possession of most observers. With smaller telescopes, we may discover an oval figure of peculiar appearance, which it would be difficult to interpret. Galileo, who first saw it in the year 1610, recognised this peculiarity, but did not know what it meant. Seeing something in the centre with two projecting arms, one on each side, he concluded that the planet was triple-shaped. This was, at the time, all he could learn respecting it, as the telescopes he possessed were very humble, compared with those now used by astronomers. The first constructed by him magnified but three times; his second, eight times; and his best, only thirty times, which is no better than a common ship-glass.

It was the practice of the astronomers of those days to give the first intimation of a new discovery in a Latin verse, the letters of which were transposed. This would enable them to claim priority, in case any other person should contest the honor of the discovery, and at the same time would afford opportunity to complete their observations, before they published a full account of them. Accordingly, Galileo announced the discovery of the singular appearance of Saturn under this disguise, in a line which, when the transposed letters were restored to their proper places, signified that he had observed, "that the most distant planet is triple-formed."[13] He shortly afterwards, at the request of his patron, the Emperor Rodolph, gave the solution, and added, "I have, with great admiration, observed that Saturn is not a single star, but three together, which, as it were, touch each other; they have no relative motion, and are constituted of this form, oOo, the middle one being somewhat larger than the two lateral ones. If we examine them with an eyeglass which magnifies the surface less than one thousand times, the three stars do not appear very distinctly, but Saturn has an oblong appearance, like that of an olive, thus,

⬭ . Now, I have discovered a court for Jupiter, (alluding to his satellites,) and two servants for this old man, (Saturn,) who aid his steps, and never quit his side."

It was by this mystic light that Galileo groped his way through an organization which, under the more powerful glasses of his successors, was to expand into a mighty orb, encompassed by splendid rings of vast dimensions, the whole attended by seven bright satellites. This system was first fully developed by Huyghens, a Dutch astronomer, about forty years afterwards.[14] It requires a superior telescope to see it to advantage; but, when seen through such a telescope, it is one of the most charming spectacles afforded to that instrument. To give some idea of the properties of a telescope suited to such observations, I annex an extract from an account, that was published a few years since, of a telescope constructed by Mr. Tully, a distinguished English artist. "The length of the instrument was twelve feet, but was easily adjusted, and was perfectly steady. The magnifying powers ranged from two hundred to seven hundred and eighty times; but the great excellence of the telescope consisted more in the superior distinctness and brilliancy with which objects were seen through it, than in its magnifying powers. With a power of two hundred and forty, the light of Jupiter was almost more than the eye could bear, and his satellites appeared as bright as Sirius, but with a clear and steady light; and the belts and spots on the face of the planet were most distinctly defined. With a power of nearly four hundred, Saturn appeared large and well defined, and was one of the most beautiful objects that can well be conceived."

That the ring is a solid opaque substance, is shown by its throwing its shadow on the body of the planet on the side nearest the sun, and on the other side receiving that of the body. The ring encompasses the equatorial regions of the planet, and the planet revolves on an axis which is perpendicular to the plane of the ring in about ten and a half hours. This is known by observing the rotation of certain dusky spots, which sometimes appear on its surface. This motion is nearly the same with the diurnal motion of Jupiter, subjecting places on the equator of the planet to a very swift revolution, and occasioning a high degree of compression at the poles, the equatorial being to the polar diameter in the high ratio of eleven to ten.

Saturn's ring, in its revolution around the sun, *always remains parallel to itself.* If we hold opposite to the eye a circular ring or disk, like a piece of coin, it will appear as a complete circle only when it is at right angles to the axis of vision. When it is oblique to that axis, it will be projected into an ellipse more and more flattened, as its obliquity is increased, until, when its plane coincides with the axis of vision, it is projected into a straight line. Please to take some circle, as a flat plate, (whose rim may well represent the ring of Saturn,) and hold it in these different positions before the eye. Now, place on the table a lamp to represent the sun, and holding the ring at a certain distance, inclined a little towards the lamp, carry it round the lamp, always keeping it parallel to itself. During its revolution, it will twice present its edge to the lamp at opposite points; and twice, at places ninety degrees distant from those points, it will present its broadest face towards the lamp. At intermediate points, it will exhibit an ellipse more or less open, according as it is nearer one or the other of the preceding positions. It will be seen, also, that in one half of the revolution, the lamp shines on one side of the ring, and in the other half of the revolution, on the other side.

Such would be the successive appearances of Saturn's ring to a spectator on the sun; and since the earth is, in respect to so distant a body as Saturn, very near the sun, these appearances are presented to us nearly in the same manner as though we viewed them from the sun. Accordingly, we sometimes see Saturn's ring under the form of a broad ellipse, which grows continually more and more acute, until it passes into a line, and we either lose sight of it, altogether, or, by the aid of the most powerful telescopes, we see it as a fine thread of light drawn across the disk, and projecting out from it on each side. As the whole revolution occupies thirty years, and the edge is presented to the sun twice in the revolution, this last phenomenon, namely, the disappearance of the ring, takes place every fifteen years.

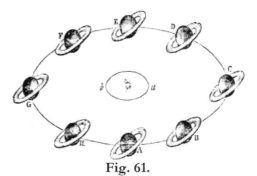

Fig. 61.

You may perhaps gain a still clearer idea of the foregoing appearances from the following diagram, Fig. 61. Let A, B, C, &c., represent successive positions of Saturn and his ring, in different parts of his orbit, while *a b* represents the orbit of the earth. Please to remark, that these orbits are drawn so elliptical, not to represent the eccentricity of either the earth's or Saturn's orbit, but merely as the projection of circles seen very obliquely. Also, imagine one half of the body of the planet and of the ring to be above the plane of the paper, and the other half below it. Were the ring, when at C and G, perpendicular to C G, it would be seen by a spectator situated at *a* or *b* as a perfect circle; but being inclined to the line of vision twenty-eight degrees four minutes, it is projected into an ellipse. This ellipse contracts in breadth as the ring passes towards its nodes at A and E, where it dwindles into a straight line. From E to G the ring opens again, becomes broadest at

G, and again contracts, till it becomes a straight line at A, and from this point expands, till it recovers its original breadth at C. These successive appearances are all exhibited to a telescope of moderate powers.

The ring is extremely *thin*, since the smallest satellite, when projected on it, more than covers it. The thickness is estimated at only one hundred miles. Saturn's ring shines wholly by *reflected light* derived from the sun. This is evident from the fact that that side only which is turned towards the sun is enlightened; and it is remarkable, that the illumination of the ring is greater than that of the planet itself, but the outer ring is less bright than the inner. Although we view Saturn's ring nearly as though we saw it from the sun, yet the plane of the ring produced may pass between the earth and the sun, in which case, also, the ring becomes invisible, the illuminated side being wholly turned from us. Thus, when the ring is approaching its node at E, a spectator at *a* would have the dark side of the ring presented to him. The ring was invisible in 1833, and will be invisible again in 1847. The northern side of the ring will be in sight until 1855, when the southern side will come into view. It appears, therefore, that there are three causes for the disappearance of Saturn's ring: first, when the edge of the ring is presented to the sun; secondly, when the edge is presented to the earth; and thirdly, when the unilluminated side is towards the earth.

Saturn's ring *revolves* in its own plane in about ten and a half hours. La Place inferred this from the doctrine of universal gravitation. He proved that such a rotation was necessary; otherwise, the matter of which the ring is composed would be precipitated upon its primary. He showed that, in order to sustain itself, its period of rotation must be equal to the time of revolution of a satellite, circulating around Saturn at a distance from it equal to that of the middle of the ring, which period would be about ten and a half hours. By means of spots in the ring, Dr. Herschel followed the ring in its rotation, and actually found its period to be the same as assigned by La Place,—a coincidence which beautifully exemplifies the harmony of truth.

Although the rings have very nearly the same centre with the planet itself, yet, recent measurements of extreme delicacy have demonstrated, that the coincidence is not mathematically exact, but that the centre of gravity of the rings describes around that of the body a very minute orbit. "This fact," says Sir J. Herschel, "unimportant as it may seem, is of the utmost consequence to the stability of the system of rings. Supposing them mathematically perfect in their circular form, and exactly concentric with the planet, it is demonstrable that they would form (in spite of their centrifugal force) a system in a state of unstable equilibrium, which the slightest external power would subvert, not by causing a rupture in the substance of the rings, but by precipitating them unbroken upon the surface of the planet." The ring may be supposed of an unequal breadth in its different parts, and as consisting of irregular solids, whose common centre of gravity does not coincide with the centre of the figure. Were it not for this distribution of matter, its equilibrium would be destroyed by the slightest force, such as the attraction of a satellite, and the ring would finally precipitate itself upon the planet. Sir J. Herschel further observes, that, "as the smallest difference of velocity between the planet and its rings must infallibly precipitate the rings upon the planet, never more to separate, it follows, either that their motions in their common orbit round the sun must have been adjusted to each other by an external power, with the minutest precision, or that the rings must have been formed about the planet while subject to their common orbitual motion, and under the full and free influence of all the acting forces.

"The rings of Saturn must present a magnificent spectacle from those regions of the planet which lie on their enlightened sides, appearing as vast arches spanning the sky from horizon to horizon, and holding an invariable situation among the stars. On the other hand, in the region beneath the dark side, a solar eclipse of fifteen years in duration, under their shadow, must afford (to our ideas) an inhospitable abode to animated beings, but ill compensated by the full light of its satellites. But we shall do wrong to judge of the fitness or unfitness of their condition, from what we see around us, when, perhaps, the very combinations which convey to our minds only images of horror, may be in reality theatres of the most striking and glorious displays of beneficent contrivance."

Saturn is attended by *seven satellites.* Although they are bodies of considerable size, their great distance prevents their being visible to any telescope but such as afford a strong light and high magnifying powers. The outermost satellite is distant from the planet more than thirty times the planet's diameter, and is by far the largest of the whole. It exhibits, like the satellites of Jupiter, periodic variations of light, which prove its revolution on its axis in the time of a sidereal revolution about Saturn, as is the case with our moon, while performing its circuit about the earth. The next satellite in order, proceeding inwards, is tolerably conspicuous; the three next are very minute, and require powerful telescopes to see them; while the two interior satellites, which just skirt the edge of the ring, and move exactly in its plane, have never been discovered but with the most powerful telescopes which human art has yet constructed, and then only under peculiar circumstances. At the time of the disappearance of the rings, (to ordinary telescopes,) they were seen by Sir William Herschel, with his great telescope, projected along the edge of the ring, and threading, like beads, the thin fibre of light to which the ring is then reduced. Owing to the obliquity of the ring, and of the orbits of the satellites, to that of their primary, there are no eclipses of the satellites, the two interior ones excepted, until near the time when the ring is seen edgewise.

"The firmament of Saturn will unquestionably present to view a more magnificent and diversified scene of celestial phenomena than that of any other planet in our system. It is placed nearly in the middle of that space which intervenes between the sun and the orbit of the remotest planet. Including its rings and satellites, it may be considered as the largest body or system of bodies within the limits of the solar system; and it excels them all in the sublime and diversified apparatus with which it is accompanied. In these respects, Saturn may justly be considered as the sovereign among the planetary hosts. The prominent parts of its celestial scenery may be considered as belonging to its own system of rings and satellites, and the views which will occasionally be opened of the firmament of the fixed stars; for few of the other planets will make their appearance in its sky. Jupiter will appear alternately as a morning and an evening star, with about the same degree of brilliancy it exhibits to us; but it will seldom be conspicuous, except near the period of its greatest elongation; and it will never appear to remove from the sun further than thirty-seven degrees, and consequently will not appear so conspicuous, nor for such a length of time, as Venus does to us. Uranus is the only other planet which will be seen from Saturn, and it will there be distinctly perceptible, like a star of the third magnitude, when near the time of its opposition to the sun. But near the time of its conjunction it will be completely invisible, being then eighteen hundred millions of miles more distant than at the opposition, and eight hundred millions of miles more distant from Saturn than it ever is from the earth at any period."[15]

URANUS.—Uranus is the remotest planet belonging to our system, and is rarely visible, except to the telescope. Although his diameter is more than four times that of the earth, being thirty-five thousand one hundred and twelve miles, yet his distance from the sun is likewise nineteen times as great as the earth's distance, or about eighteen hundred millions of miles. His revolution around the sun occupies nearly eighty-four years, so that his position in the heavens, for several years in succession, is nearly stationary. His path lies very nearly in the ecliptic, being inclined to it less than one degree. The sun himself, when seen from Uranus dwindles almost to a star, subtending, as it does, an angle of only one minute and forty seconds; so that the surface of the sun would appear there four hundred times less than it does to us. This planet was discovered by Sir William Herschel on the thirteenth of March, 1781. His attention was attracted to it by the largeness of its disk in the telescope; and finding that it shifted its place among the stars, he at first took it for a comet, but soon perceived that its orbit was not eccentric, like the orbits of comets, but nearly circular, like those of the planets. It was then recognised as a new member of the planetary system, a conclusion which has been justified by all succeeding observations. It was named by the discoverer the *George Star*, (Georgium Sidus,) after his munificent patron, George the Third; in the United States, and in some other countries, it was called *Herschel*; but the name *Uranus*, from a Greek word, (Ουρανος, *Ouranos*,) signifying the oldest of the gods, has finally prevailed. So distant is Uranus from the sun, that light itself, which moves nearly twelve millions of miles every minute, would require more than two hours and a half to pass to it from the sun.

And now, having contemplated all the planets separately, just cast your eyes on the diagram facing page 236, Fig. 53, and you will see a comparative view of the various magnitudes of the sun, as seen from each of the planets.

Uranus is attended by *six satellites*. So minute objects are they, that they can be seen only by powerful telescopes. Indeed, the existence of more than two is still considered as somewhat doubtful. These two, however, offer remarkable and indeed quite unexpected and unexampled peculiarities. Contrary to the unbroken analogy of the whole planetary system, *the planes of their orbits are nearly perpendicular to the ecliptic*, and in these orbits their motions are retrograde; that is, instead of advancing from west to east around their primary, as is the case with all the other planets and satellites, they move in the opposite direction. With this exception, all the motions of the planets, whether around their own axes, or around the sun, are from west to east. The sun himself turns on his axis from west to east; all the primary planets revolve around the sun from west to east; their revolutions on their own axes are also in the same direction; all the secondaries, with the single exception above mentioned, move about their primaries from west to east; and, finally, such of the secondaries as have been discovered to have a diurnal revolution, follow the same course. Such uniformity among so many motions could have resulted only from forces impressed upon them by the same Omnipotent hand; and few things in the creation more distinctly proclaim that God made the world.

Retiring now to this furthest verge of the solar system, let us for a moment glance at the aspect of the firmament by night. Notwithstanding we have taken a flight of eighteen hundred millions of miles, the same starry canopy bends over our heads; Sirius still shines with exactly the same splendor as here; Orion, the Scorpion, the Great and the Little Bear, all occupy the same stations; and the Galaxy spans the sky with the same soft and mysterious light. The planets, however, with the exception of Saturn, are all lost to the view, being too near the sun ever to be seen; and Saturn himself is visible only at distant intervals, at periods of fifteen years, when at its greatest elongations from the sun, and is then too near the sun to permit a clear view of his rings, much less of the satellites that unite with the rings to compose his gorgeous retinue. Comets, moving slowly as they do when so distant from the sun, will linger much longer in the firmament of Uranus than in ours.

Although the sun sheds by day a dim and cheerless light, yet the six satellites that enlighten and diversify the nocturnal sky present interesting aspects. "Let us suppose one satellite presenting a surface in the sky eight or ten times larger than our moon; a second, five or six times larger; a third, three times larger; a fourth, twice as large; a fifth, about the same size as the moon; a sixth, somewhat smaller; and, perhaps, three or four others of different apparent dimensions: let us suppose two or three of those, of different phases, moving along the concave of the sky, at one period four or five of them dispersed through the heavens, one rising above the horizon, one setting, one on the meridian, one towards the north, and another towards the south; at another period, five or six of them displaying their lustre in the form of a half moon, or a crescent, in one quarter of the heavens; and, at another time, the whole of these moons shining, with full enlightened hemispheres, in one glorious assemblage, and we shall have a faint idea of the beauty, variety, and sublimity of the firmament of Uranus."[16]

The New Planets,—Ceres, Pallas, Juno, and Vesta.—The commencement of the present century was rendered memorable in the annals of astronomy, by the discovery of four new planets, occupying the long vacant tract between Mars and Jupiter. Kepler, from some analogy which he found to subsist among the distances of the planets from the sun, had long before suspected the existence of one at this distance; and his conjecture was rendered more probable by the discovery of Uranus, which follows the analogy of the other planets. So strongly, indeed, were astronomers impressed with the idea that a planet would be found between Mars and Jupiter, that, in the hope of discovering it, an association was formed on the continent of Europe, of twenty-four observers, who divided the sky into as many zones, one of which was allotted to each member of the association. The discovery of the first of these bodies was, however, made accidentally by Piazzi, an astronomer of Palermo, on the first of January, 1801. It was shortly afterwards lost sight of on account of its proximity to the sun, and was not seen again until the close of the year, when it was re-discovered in Germany. Piazzi called it *Ceres*, in honor of the tutelary goddess of Sicily, and her emblem, the sickle, (?) has been adopted as its appropriate symbol.

The difficulty of finding Ceres induced Dr. Olbers, of Bremen, to examine with particular care all the small stars that lie near her path, as seen from the earth; and, while prosecuting these observations, in March, 1802, he discovered another similar body, very nearly at the same distance from the sun, and resembling the former in many other particulars. The discoverer gave to this second planet the name of *Pallas*, choosing for its symbol the lance, (⚢) the characteristic of Minerva.

The most surprising circumstance connected with the discovery of *Pallas* was the existence of two planets at nearly the same distance from the sun, and apparently crossing the ecliptic in the same part of the heavens, or having the same node. On account of this singularity, Dr. Olbers was led to conjecture that Ceres and Pallas are only fragments of a larger planet, which had formerly circulated at the same distance, and been shattered by some internal convulsion. The hypothesis suggested the probability that there might be other fragments, whose orbits might be expected to cross the ecliptic at a common point, or to have the same node. Dr. Olbers, therefore, proposed to examine carefully, every month, the two opposite parts of the heavens in which the orbits of Ceres and Pallas intersect one another, with a view to the discovery of other planets, which might be sought for in those parts with a greater chance of success, than in a wider zone, embracing the entire limits of these orbits. Accordingly, in 1804, near one of the nodes of Ceres and Pallas, a third planet was discovered. This was called *Juno*, and the character (⚵) was adopted for its symbol, representing the starry sceptre of the Queen of Olympus. Pursuing the same researches, in 1807 a fourth planet was discovered, to which was given the name of *Vesta*, and for its symbol the character (⚶) was chosen,—an altar surmounted with a censer holding the sacred fire.

The *average distance* of these bodies from the sun is two hundred and sixty-one millions of miles; and it is remarkable that their orbits are very near together. Taking the distance of the earth from the sun for unity, their respective distances are 2.77, 2.77, 2.67, 2.37. Their *times* of revolution around the sun are nearly equal, averaging about four and a half years.

In respect to the *inclination of their orbits* to the ecliptic, there is also considerable diversity. The orbit of Vesta is inclined only about seven degrees, while that of Pallas is more than thirty-four degrees. They all, therefore, have a higher inclination than the orbits of the old planets, and of course make excursions from the ecliptic beyond the limits of the zodiac. Hence they have been called the *ultra-zodiacal planets*. When first discovered, before their place in the system was fully ascertained it was also proposed to call them *asteroids*, a name implying that they were planets under the form of stars. Their title, however, to take their rank among the primary planets, is now generally conceded.

The *eccentricity of their orbits* is also, in general, greater than that of the old planets. You will recollect that this language denotes that their orbits are more elliptical, or depart further from the circular form. The eccentricities of the orbits of Pallas and Juno exceed that of the orbit of Mercury. The asteroids differ so much, however, in eccentricity, that their orbits may cross each other. The orbits of the old planets are so nearly circular, and at such a great distance apart, that there is no danger of their interfering with each other. The earth, for example, when at its nearest distance from the sun, will never come so near it as Venus is when at its greatest distance, and therefore can never cross the orbit of Venus. But since the average distance of Ceres and Pallas from the sun is about the same, while the eccentricity of the orbit of Pallas is much greater than that of Ceres, consequently, Pallas may come so near to the sun at its perihelion, as to cross the orbit of Ceres.

The *small size* of the asteroids constitutes one of their most remarkable peculiarities. The difficulty of estimating the apparent diameter of bodies at once so very small and so far off, would lead us to expect different results in the actual estimates. Accordingly, while Dr. Herschel estimates the diameter of Pallas at only eighty miles, Schroeter places it as high as two thousand miles, or about the diameter of the moon. The volume of Vesta is estimated at only one fifteen thousandth part of the earth's, and her surface is only about equal to that of the kingdom of Spain.

These little bodies are surrounded by *atmospheres* of great extent, some of which are uncommonly luminous, and others appear to consist of nebulous matter, like that of comets. These planets shine with a more vivid light than might be expected, from their great distance and diminutive size; but a good telescope is essential for obtaining a distinct view of their phenomena.

Although the great chasm which occurs between Mars and Jupiter,—a chasm of more than three hundred millions of miles,—suggested long ago the idea of other planetary bodies occupying that part of the solar system, yet the discovery of the asteroids does not entirely satisfy expectation since they are bodies so dissimilar to the other members of the series in size, in appearance, and in the form and inclinations of their orbits. Hence, Dr. Olbers, the discoverer of three of these bodies, held that they were fragments of a single large planet, which once occupied that place in the system, and which exploded in different directions by some internal violence. Of the fragments thus projected into space, some would be propelled in such directions and with such velocities, as, under the force of projection and that of the solar attraction would make them revolve in regular orbits around the sun. Others would be so projected among the other bodies in the system, as to be thrown in very irregular orbits, apparently wandering lawless through the skies. The larger fragments would receive the least impetus from the explosive force, and would therefore circulate in an orbit deviating less than any other of the fragments from the original path of the large planet; while the lesser fragments, being thrown off with greater velocity, would revolve in orbits more eccentric, and more inclined to the ecliptic.

Dr. Brewster, editor of the 'Edinburgh Encyclopedia,' and the well-known author of various philosophical works, espoused this hypothesis with much zeal; and, after summing up the evidence in favor of it, he remarks as follows: "These singular resemblances in the motions of the greater fragments, and in those of the lesser fragments, and the striking coincidences between theory and observation in the eccentricity of their orbits, in their inclination to the ecliptic, in the position of their nodes, and in the places of their perihelia, are phenomena which could not possibly result from chance, and which concur to prove, with an evidence amounting almost to demonstration, that the four new planets have diverged from one common node, and have therefore composed a single planet."

The same distinguished writer supposes that some of the smallest fragments might even have come within reach of the earth's attraction, and by the combined effects of their projectile forces and the attraction of the earth, be made to revolve around this body as the larger fragments are carried around the sun; and that these are in fact the bodies which afford those *meteoric stones* which are occasionally precipitated to the earth. It is now a well-ascertained fact, a fact which has been many times verified in our own country, that large meteors, in the shape of fire-balls, traversing the atmosphere, sometimes project to the earth masses of stony or ferruginous matter. Such were the meteoric stones which fell at Weston, in Connecticut, in 1807, of which a full and interesting account was published, after a minute examination of the facts, by Professors Silliman and Kingsley, of Yale College. Various accounts of similar occurrences may be found in different volumes of the American Journal of Science. It is for the production of these wonderful phenomena that Dr. Brewster supposes the explosion of the planet, which, according to Dr. Olbers, produced the asteroids, accounts. Others, however, as Sir John Herschel, have been disposed to ascribe very little weight to the doctrine of Olbers.

LETTER XXIV.

THE PLANETARY MOTIONS.——KEPLER'S LAWS.——KEPLER.

"God of the rolling orbs above!
Thy name is written clearly bright
 In the warm day's unvarying blaze,
Or evening's golden shower of light;
For every fire that fronts the sun,
And every spark that walks alone
Around the utmost verge of heaven,
Was kindled at thy burning throne."—*Peabody*.

IF we could stand upon the sun and view the planetary motions, they would appear to us as simple as the motions of equestrians riding with different degrees of speed around a large ring, of which we occupied the centre. We should see all the planets coursing each other from west to east, through the same great highway, (the Zodiac,) no one of them, with the exception of the asteroids, deviating more than seven degrees from the path pursued by the earth. Most of them, indeed, would always be seen moving much nearer than that to the ecliptic. We should see the planets moving on their way with various degrees of speed. Mercury would make the entire circuit in about three months, hurrying on his course with a speed about one third as great as that by which the moon revolves around the earth. The most distant planets, on the other hand, move at so slow a pace, that we should see Mercury, Venus, the Earth, and Mars, severally overtaking them a great many times, before they had completed their revolutions. But though the movements of some were comparatively rapid, and of others extremely slow, yet they would not seem to differ materially, in other respects: each would be making a steady and nearly uniform march along the celestial vault.

Such would be the simple and harmonious motions of the planets, as they would be seen from the sun, the centre of their motions; and such they are, in fact. But two circumstances conspire to make them appear exceedingly different from these, and vastly more complicated; one is, that we view them out of the centre of their motions; the other, that we are ourselves in motion. I have already explained to you the effect which these two causes produce on the apparent motions of the inferior planets, Mercury and Venus. Let us now see how they effect those of the superior planets, Mars, Jupiter, Saturn, and Uranus.

Orreries, or machines intended to exhibit a model of the solar system, are sometimes employed to give a popular view of the planetary motions; but they oftener mislead than give correct ideas. They may assist reflection, but they can never supply its place. The impossibility of representing things in their just proportions will be evident, when we reflect that, to do this, if in an orrery we make Mercury as large as a cherry, we should have to represent the sun six feet in diameter. If we preserve the same proportions, in regard to distance, we must place Mercury two hundred and fifty feet, and Uranus twelve thousand five hundred feet, or more than two miles from the sun. The mind of the student of astronomy must, therefore, raise itself from such imperfect representations of celestial phenomena, as are afforded by artificial mechanism, and, transferring his contemplations to the celestial regions themselves, he must conceive of the sun and planets as bodies that bear an insignificant ratio to the immense spaces in which they circulate, resembling more a few little birds flying in the open sky, than they do the crowded machinery of an orrery.

The *real* motions of the planets, indeed, or such as orreries usually exhibit, are very easily conceived of, as before explained; but the *apparent* motions are, for the most part, entirely different from these. The apparent motions of the inferior planets have been already explained. You will recollect that Mercury and Venus move backwards and forwards across the sun, the former never being seen further than twenty-nine degrees, and the latter never more than about forty-seven degrees, from that luminary; that, while passing from the greatest elongation on one side, to the greatest elongation on the other side, through the superior conjunction, the apparent motions of these planets are accelerated by the motion of the earth; but that, while moving through the inferior conjunction, at which time their motions are retrograde, they are apparently retarded by the earth's motion. Let us now see what are the apparent motions of the superior planets.

Let A, B, C, Fig. 62, page 294, represent the earth in different positions in its orbit, M, a superior planet, as Mars, and N R, an arc of the concave sphere of the heavens. First, suppose the planet to remain at rest in M, and let us see what apparent motions it will receive from the real motions of the earth. When the earth is at B, it will see the planet in the heavens at N; and as the earth moves successively through C, D, E, F, the planet will appear to move through O, P, Q, R. B and F are the two points of greatest elongation of the earth from the sun, as seen from the planet; hence, between these two points, while passing through its orbit most remote from the planet, (when the planet is seen in superior conjunction,) the earth, by its own motion, gives an apparent motion to the planet in the order of the signs; that is, the *apparent* motion given by the *real* motion of the earth is *direct*. But in passing from F to B through A, when the planet is seen in opposition, the apparent motion given to the planet by the earth's motion is from R to N, and is therefore *retrograde*. As the arc described by the earth, when the motion is direct, is much greater than when the motion is retrograde, while the apparent arc of the heavens described by the planet from N to R, in the one case, and from R to N, in the other, is the same in both cases, the retrograde motion is much swifter than the direct, being performed in much less time.

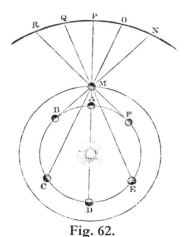

Fig. 62.

But the superior planets are not in fact at rest, as we have supposed, but are all the while moving eastward, though with a slower motion than the earth. Indeed, with respect to the remotest planets, as Saturn and Uranus, the forward motion is so exceedingly slow, that the above representation is nearly true for a single year. Still, the effect of the real motions of all the superior planets, eastward, is to increase the direct apparent motion communicated by the earth, and to diminish the retrograde motion. This will be evident from inspecting the figure; for if the planet *actually* moves eastward while it is *apparently* carried eastward by the earth's motion, the whole motion eastward will be equal to the sum of the two; and if, while it is really moving eastward, it is apparently carried westward still more by the earth's motion, the retrograde movement will equal the difference of the two.

If Mars stood still while the earth went round the sun, then a second opposition, as at A, would occur at the end of one year from the first; but, while the earth is performing this circuit, Mars is also moving the same way, more than half as fast; so that, when the earth returns to A, the planet has already performed more than half the same circuit, and will have completed its whole revolution before the earth comes up with it. Indeed Mars, after having been seen once in opposition, does not come into opposition again until after two years and fifty days. And since the planet is then comparatively near to us, as at M, while the earth is at A, and appears very large and bright, rising unexpectedly about the time the sun sets, he surprises the world as though it were some new celestial body. But on account of the slow progress of Saturn and Uranus, we find, after having performed one circuit around the sun, that they are but little advanced beyond where we left them at the last opposition. The time between one opposition of Saturn and another is only a year and thirteen days.

It appears, therefore, that the superior planets steadily pursue their course around the sun, but that their apparent retrograde motion, when in opposition, is occasioned by our passing by them with a swifter motion, of which we are unconscious, like the apparent backward motion of a vessel, when we overtake it and pass by it rapidly in a steam-boat.

Such are the real and the apparent motions of the planets. Let us now turn our attention to the *laws of the planetary orbits*.

There are three great principles, according to which the motions of the earth and all the planets around the sun are regulated, called KEPLER'S LAWS, having been first discovered by the astronomer whose name they bear. They may appear to you, at first, dry and obscure; yet they will be easily understood from the explanations which follow; and so important have they proved in astronomical inquiries, that they have acquired for their renowned discoverer the appellation of the '*Legislator of the Skies*.' We will consider each of these laws separately; and, for the sake of rendering the explanation clear and intelligible, I shall perhaps repeat some things that have been briefly mentioned before.

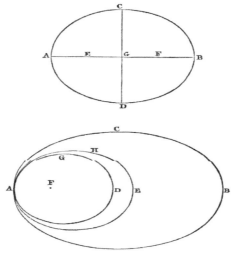

Fig. 63. Fig. 64.

FIRST LAW.—*The orbits of the earth and all the planets are ellipses, having the sun in the common focus.* In a circle, all the diameters are equal to one another; but if we take a metallic wire or hoop, and draw it out on opposite sides, we elongate it into an ellipse, of which the different diameters are very unequal. That which connects the points most distant from each other is called the *transverse*, and that which is at right angles to this is called the *conjugate*, axis. Thus, A B, Fig. 63, is the transverse axis, and C D, the conjugate of the ellipse A B C. By such a process of elongating the circle into an ellipse, the centre of the circle may be conceived of as drawn opposite ways to E and F, each of which becomes a *focus*, and both together are called the *foci* of the ellipse. The distance G E, or G F, of the focus from the centre is called the *eccentricity* of the ellipse; and the ellipse is said to be more or less eccentric, as the distance of the focus from the centre is greater or less. Figure 64 represents such a collection of ellipses around the common focus F, the innermost, A G D, having a small eccentricity, or varying little from a circle, while the outermost, A C B, is an eccentric ellipse. The orbits of all the bodies that revolve about the sun, both planets and comets, have, in like manner, a common focus, in which the sun is situated, but they differ in eccentricity. Most of the planets have orbits of very little eccentricity, differing little from circles, but comets move in very eccentric ellipses. The earth's path around the sun varies so little from a circle, that a diagram representing it truly would scarcely be distinguished from a perfect circle; yet, when the comparative distances of the sun from the earth are taken at different seasons of the year, we find that the difference between their greatest and least distances is no less than three millions of miles.

SECOND LAW.—*The radius vector of the earth, or of any planet, describes equal areas in equal times.* You will recollect that the radius vector is a line drawn from the centre of the sun to a planet revolving about the sun. This definition I have somewhere given you before, and perhaps it may appear to you like needless repetition to state it again. In a book designed for systematic instruction, where all the articles are distinctly numbered, it is commonly sufficient to make a reference back to the article where the point in question is explained; but I think, in Letters like these, you will bear with a little repetition, rather than be at the trouble of turning to the Index and hunting up a definition long since given.

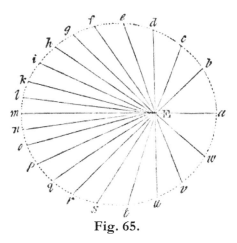

Fig. 65.

In Figure 65, *E a*, *E b*, *E c*, &c., are successive representations of the radius vector. Now, if a planet sets out from *a*, and travels round the sun in the direction of *a b c*, it will move faster when nearer the sun, as at *a*, than when more remote from it, as at *m*; yet, if *a b* and *m n* be arcs described in equal times, then, according to the foregoing law, the space *E a b* will be equal to the space *E m n*; and the same is true of all the other spaces described in equal times. Although the figure *E a b* is much shorter than *E m n*, yet its greater breadth exactly counterbalances the greater length of those figures which are described by the radius vector where it is longer.

THIRD LAW.—*The squares of the periodical times are as the cubes of the mean distances from the sun.* The periodical time of a body is the time it takes to complete its orbit, in its revolution about the sun. Thus the earth's periodic time is one year, and that of the planet Jupiter about twelve years. As Jupiter takes so much longer time to travel round the sun than the earth does, we might suspect that his orbit is larger than that of the earth, and of course, that he is at a greater distance from the sun; and our first thought might be, that he is probably twelve times as far off; but Kepler discovered that the distance does not increase as fast as the times increase, but that the planets which are more distant from the sun actually move slower than those which are nearer. After trying a great many proportions, he at length found that, if we take the squares of the periodic times of two planets, the greater square contains the less, just as often as the cube of the distance of the greater contains that of the less. This fact is expressed by saying, that the squares of the periodic times are to one another as the cubes of the distances.

This law is of great use in determining the distance of the planets from the sun. Suppose, for example, that we wish to find the distance of Jupiter. We can easily determine, from observation, what is Jupiter's periodical time, for we can actually see how long it takes for Jupiter, after leaving a certain part of the heavens to come round to the same part again. Let this period be twelve years. The earth's period is of course one year; and the distance of the earth, as determined from the sun's horizontal parallax, as already explained, is about ninety-five millions of miles. Now, we have here three terms of a proportion to find the fourth, and therefore the solution is merely a simple case of the rule of three. Thus:—the square of 1 year : square of 12 years : cube of 95,000,000 : cube of Jupiter's distance. The three first terms being known, we have only to multiply together the second and third and divide by the first, to obtain the fourth term, which will give us the cube of Jupiter's distance from the sun; and by extracting the cube root of this sum, we obtain the distance itself. In the same manner we may obtain the respective distances of all the other planets.

So truly is this a law of the solar system, that it holds good in respect to the new planets, which have been discovered since Kepler's time, as well as in the case of the old planets. It also holds good in respect to comets, and to all bodies belonging to the solar system, which revolve around the sun as their centre of motion. Hence, it is justly regarded as one of the most interesting and important principles yet developed in astronomy.

But who was this Kepler, that gained such an extraordinary insight into the laws of the planetary system, as to be called the 'Legislator of the Skies?' John Kepler was one of the most remarkable of the human race, and I think I cannot gratify or instruct you more, than by occupying the remainder of this Letter with some particulars of his history.

Kepler was a native of Germany. He was born in the Duchy of Wurtemberg, in 1571. As Copernicus, Tycho Brahe, Galileo, Kepler, and Newton, are names that are much associated in the history of astronomy, let us see how they stood related to each other in point of time. Copernicus was born in 1473; Tycho, in 1546; Galileo, in 1564; Kepler, in 1571; and

Newton, in 1642. Hence, Copernicus was seventy-three years before Tycho, and Tycho ninety-six years before Newton. They all lived to an advanced age, so that Tycho, Galileo, and Kepler, were contemporary for many years; and Newton, as I mentioned in the sketch I gave you of his life, was born the year that Galileo died.

Kepler was born of parents who were then in humble circumstances, although of noble descent. Their misfortunes, which had reduced them to poverty, seem to have been aggravated by their own unhappy dispositions; for his biographer informs us, that "his mother was treated with a degree of barbarity by her husband and brother-in-law, that was hardly exceeded by her own perverseness." It is fortunate, therefore, that Kepler, in his childhood, was removed from the immediate society and example of his parents, and educated at a public school at the expense of the Duke of Wurtemberg. He early imbibed a taste for natural philosophy, but had conceived a strong prejudice against astronomy, and even a contempt for it, inspired, probably, by the arrogant and ridiculous pretensions of the astrologers, who constituted the principal astronomers of his country. A vacant post, however, of teacher of astronomy, occurred when he was of a suitable age to fill it, and he was compelled to take it by the authority of his tutors, though with many protestations, on his part, wishing to be provided for in some other more brilliant profession.

Happy is genius, when it lights on a profession entirely consonant to its powers, where the objects successively presented to it are so exactly suited to its nature, that it clings to them as the loadstone to its kindred metal among piles of foreign ores. Nothing could have been more congenial to the very mental constitution of Kepler, than the study of astronomy,—a science where the most capacious understanding may find scope in unison with the most fervid imagination.

Much as has been said against hypotheses in philosophy, it is nevertheless a fact, that some of the greatest truths have been discovered in the pursuit of hypotheses, in themselves entirely false; truths, moreover, far more important than those assumed by the hypotheses; as Columbus, in searching for a northwest passage to India, discovered a new world. Thus Kepler groped his way through many false and absurd suppositions, to some of the most sublime discoveries ever made by man. The fundamental principle which guided him was not, however, either false or absurd. It was, that God, who made the world, had established, throughout all his works, fixed laws,—laws that are often so definite as to be capable of expression in exact numerical terms. In accordance with these views, he sought for numerical relations in the disposition and arrangement of the planets, in respect to their number, the times of their revolution, and their distances from one another. Many, indeed, of the subordinate suppositions which he made, were extremely fanciful; but he tried his own hypotheses by a rigorous mathematical test, wherever he could apply it; and as soon as he discovered that a supposition would not abide this test, he abandoned it without the least hesitation, and adopted others, which he submitted to the same severe trial, to share, perhaps, the same fate. "After many failures," he says, "I was comforted by observing that the motions, in every case, seemed to be connected with the distances; and that, when there was a great gap between the orbits, there was the same between the motions. And I reasoned that, if God had adapted motions to the orbits in some relation to the distances, he had also arranged the distances themselves in relation to something else."

In two years after he commenced the study of astronomy, he published a book, called the '*Mysterium Cosmographicum*,' a name which implies an explanation of the mysteries involved in the construction of the universe. This work was full of the wildest speculations and most extravagant hypotheses, the most remarkable of which was, that the distances of the planets from the sun are regulated by the relations which subsist between the five regular solids. It is well known to geometers, that there are and can be only five *regular solids*. These are, first, the *tetraedron*, a four-sided figure, all whose sides are equal and similar triangles; secondly, the *cube*, contained by six equal squares; thirdly, an *octaedron*, an eight-sided figure, consisting of two four-sided pyramids joined at their bases; fourthly, a *dodecaedron*, having twelve five-sided or pentagonal faces; and, fifthly, an *icosaedron*, contained by twenty equal and similar triangles. You will be much at a loss, I think, to imagine what relation Kepler could trace between these strange figures and the distances of the several planets from the sun. He thought he discovered a connexion between those distances and the spaces which figures of this kind would occupy, if interposed in certain ways between them. Thus, he says the Earth is a circle, the measure of all; round it describe a dodecaedron, and the circle including this will be the orbit of Mars. Round this circle describe a tetraedron, and the circle including this will be the orbit of Jupiter. Describe a cube round this, and the circle including it will be the orbit of Saturn. Now, inscribe in the earth an icosaedron, and the circle included in this will give the orbit of Venus. In this inscribe an octaedron, and the circle included in this will be the orbit of Mercury. On this supposed discovery Kepler exults in the most enthusiastic expressions. "The intense pleasure I have received from this discovery never can be told in words. I regretted no more time wasted; I tired of no labor; I shunned no toil of reckoning; days and nights I spent in calculations, until I could see whether this opinion would agree with the orbits of Copernicus, or whether my joy was to vanish into air. I willingly subjoin that sentiment of

Archytas, as given by Cicero; 'If I could mount up into heaven, and thoroughly perceive the nature of the world and the beauty of the stars, that admiration would be without a charm for me, unless I had some one like you, reader, candid, attentive, and eager for knowledge, to whom to describe it.' If you acknowledge this feeling, and are candid, you will refrain from blame, such as, not without cause, I anticipate; but if, leaving that to itself, you fear, lest these things be not ascertained, and that I have shouted triumph before victory, at least approach these pages, and learn the matter in consideration: you will not find, as just now, new and unknown planets interposed; that boldness of mine is not approved; but those old ones very little loosened, and so furnished by the interposition (however absurd you may think it) of rectilinear figures, that in future you may give a reason to the rustics, when they ask for the hooks which keep the skies from falling."

When Tycho Brahe, who had then retired from his famous Uraniburg, and was settled in Prague, met with this work of Kepler's, he immediately recognised under this fantastic garb the lineaments of a great astronomer. He needed such an unwearied and patient calculator as he perceived Kepler to be, to aid him in his labors, in order that he might devote himself more unreservedly to the taking of observations,—an employment in which he delighted, and in which, as I mentioned, in giving you a sketch of his history, he excelled all men of that and preceding ages. Kepler, therefore, at the express invitation of Tycho, went to Prague, and joined him in the capacity of assistant. Had Tycho been of a nature less truly noble, he might have looked with contempt on one who had made so few observations, and indulged so much in wild speculation; or he might have been jealous of a rising genius, in which he descried so many signs of future eminence as an astronomer; but, superior to all the baser motives, he extends to the young aspirant the hand of encouragement, in the following kind invitation: "Come not as a stranger, but as a very welcome friend; come, and share in my observations, with such instruments as I have with me."

Several years previous to this, Kepler, after one or two unsuccessful trials, had found him a wife, from whom he expected a considerable fortune; but in this he was disappointed; and so poor was he, that, when on his journey to Prague, in company with his wife, being taken sick, he was unable to defray the expenses of the journey, and was forced to cast himself on the bounty of Tycho.

In the course of the following year, while absent from Prague, he fancied that Tycho had injured him, and accordingly addressed to the noble Dane a letter full of insults and reproaches. A mild reply from Tycho opened the eyes of Kepler to his own ingratitude. His better feelings soon returned, and he sent to his great patron this humble apology: "Most noble Tycho!

How shall I enumerate, or rightly estimate, your benefits conferred on me! For two months you have liberally and gratuitously maintained me, and my whole family; you have provided for all my wishes; you have done me every possible kindness; you have communicated to me every thing you hold most dear; no one, by word or deed, has intentionally injured me in any thing; in short, not to your own children, your wife, or yourself, have you shown more indulgence than to me. This being so, as I am anxious to put upon record, I cannot reflect, without consternation, that I should have been so given up by God to my own intemperance, as to shut my eyes on all these benefits; that, instead of modest and respectful gratitude, I should indulge for three weeks in continual moroseness towards all your family, and in headlong passion and the utmost insolence towards yourself, who possess so many claims on my veneration, from your noble family, your extraordinary learning, and distinguished reputation. Whatever I have said or written against the person, the fame, the honor, and the learning, of your Excellency; or whatever, in any other way, I have injuriously spoken or written, (if they admit no other more favorable interpretation,) as to my grief I have spoken and written many things, and more than I can remember; all and every thing I recant, and freely and honestly declare and profess to be groundless, false, and incapable of proof." This was ample satisfaction to the generous Tycho.

"To err is human: to forgive, divine."

On Kepler's return to Prague, he was presented to the Emperor by Tycho, and honored with the title of Imperial Mathematician. This was in 1601, when he was thirty years of age. Tycho died shortly after, and Kepler succeeded him as principal mathematician to the Emperor; but his salary was badly paid, and he suffered much from pecuniary embarrassments. Although he held the astrologers, or those who told fortunes by the stars, in great contempt, yet he entertained notions of his own, on the same subject, quite as extravagant, and practised the art of casting nativities, to eke out a support for his family.

When Galileo began to observe with his telescope, and announced, in rapid succession, his wonderful discoveries, Kepler entered into them with his characteristic enthusiasm, although they subverted many of his favorite hypotheses. But such was his love of truth, that he was among the first to congratulate Galileo, and a most engaging correspondence was carried on between these master-spirits.

The first planet, which occupied the particular attention of Kepler, was Mars, the long and assiduous study of whose motions conducted him at length to the discovery of those great principles called 'Kepler's Laws.'

Rarely do we meet with so remarkable a union of a vivid fancy with a profound intellect. The hasty and extravagant suggestions of the former were submitted to the most laborious calculations, some of which, that were of great length, he repeated seventy times. This exuberance of fancy frequently appears in his style of writing, which occasionally assumes a tone ludicrously figurative. He seems constantly to contemplate Mars as a valiant hero, who had hitherto proved invincible, and who would often elude his own efforts to conquer him, "While thus triumphing over Mars, and preparing for him, as for one altogether vanquished, tabular prisons, and equated, eccentric fetters, it is buzzed here and there, that the victory is vain, and that the war is raging anew as violently as before. For the enemy, left at home a despised captive, has burst all the chains of the equation, and broken forth of the prisons of the tables. Skirmishes routed my forces of physical causes, and, shaking off the yoke, regained their liberty. And now, there was little to prevent the fugitive enemy from effecting a junction with his own rebellious supporters, and reducing me to despair, had I not suddenly sent into the field a reserve of new physical reasonings, on the rout and dispersion of the veterans, and diligently followed, without allowing the slightest respite, in the direction in which he had broken out."

But he pursued this warfare with the planet until he gained a full conquest, by the discovery of the first two of his laws, namely, that *he revolves in an elliptical orbit*, and that *his radius vector passes over equal spaces in equal times*.

Domestic troubles, however, involved him in the deepest affliction. Poverty, the loss of a promising and favorite son, the death of his wife, after a long illness;—these were some of the misfortunes that clustered around him. Although his first marriage had been an unhappy one, it was not consonant to his genius to surrender any thing with only a single trial. Accordingly, it was not long before he endeavored to repair his loss by a second alliance. He commissioned a number of his friends to look out for him, and he soon obtained a tabular list of eleven ladies, among whom his affections wavered. The progress of his courtship is thus narrated in the interesting 'Life' contained in the 'Library of Useful Knowledge.' It furnishes so fine a specimen of his eccentricities, that I cannot deny myself the pleasure of transcribing the passage for your perusal. It is taken from an account which Kepler himself gave in a letter to a friend.

"The first on the list was a widow, an intimate friend of his first wife and who, on many accounts, appeared a most eligible match. At first, she seemed favorably inclined to the proposal: it is certain that she took time to consider it, but at last she very quietly excused herself. Finding her afterwards less agreeable in person than he had anticipated, he considered it a fortunate escape, mentioning, among other objections, that she had two marriageable daughters, whom, by the way, he had got on his list for examination. He was much troubled to reconcile his astrology with the fact of his having taken so much pains about a negotiation not destined to succeed. He examined the case professionally. 'Have the stars,' says he, 'exercised any influence here? For, just about this time, the direction of the mid-heaven is in hot opposition to Mars, and the passage of Saturn through the ascending point of the zodiac, in the scheme of my nativity, will happen again next November and December. But, if these are the causes, how do they act? Is that explanation the true one, which I have elsewhere given? For I can never think of handing over to the stars the office of deities, to produce effects. Let us, therefore, suppose it accounted for by the stars, that at this season I am violent in my temper and affections, in rashness of belief, in a show of pitiful tender-heartedness, in catching at reputation by new and paradoxical notions, and the singularity of my actions; in busily inquiring into, and weighing, and discussing, various reasons; in the uneasiness of my mind, with respect to my choice. I thank God, that that did not happen which might have happened; that this marriage did not take place. Now for the others.' Of these, one was too old; another, in bad health; another, too proud of her birth and quarterings; a fourth had learned nothing but showy accomplishments, not at all suitable to the kind of life she would have to lead with him. Another grew impatient, and married a more decided admirer while he was hesitating. 'The mischief,' says he, 'in all these attachments was, that, whilst I was delaying, comparing, and balancing, conflicting reasons, every day saw me inflamed with a new passion.' By the time he reached No. 8, of his list, he found his match in this respect. 'Fortune has avenged herself at length on my doubtful inclinations. At first, she was quite complying, and her friends also. Presently, whether she did or did not consent, not only I, but she herself, did not know. After the lapse of a few days, came a renewed promise, which, however, had to be confirmed a third time: and, four days after that, she again repented her conformation, and begged to be excused from it. Upon this, I gave her up, and this time all my counsellors were of one opinion.' This was the longest courtship in the list, having lasted three whole months; and, quite disheartened by its bad success, Kepler's next attempt was of a more timid complexion. His advances to No. 9 were made by confiding to her the whole story of his recent disappointment, prudently determining to be guided in his behavior, by observing whether the treatment he experienced met with a proper degree of sympathy. Apparently, the experiment did not succeed; and, when almost reduced to despair, Kepler betook himself to the advice of a friend, who had for some time past complained that she was not consulted in this difficult negotiation. When she produced No. 10, and the first visit was paid, the report upon her was as follows: 'She has, undoubtedly, a good fortune, is of good family, and of economical habits: but her physiognomy is most horribly ugly; she would be stared at in the streets, not to mention the striking disproportion in our figures. I am lank, lean, and spare; she is short and thick. In a family notorious for fatness, she is considered superfluously fat.' The only objection to No. 11 seems to have been, her excessive youth; and when this treaty was broken off, on that account, Kepler turned his back upon all his advisers, and chose for himself one who had figured as No. 5, in his list, to whom he professes to have felt attached throughout, but from whom the representations of his friends had hitherto detained him, probably on account of her humble station."

Having thus settled his domestic affairs, Kepler now betook himself, with his usual industry, to his astronomical studies, and brought before the world the most celebrated of his publications, entitled 'Harmonics.' In the fifth book of this work he announced his *Third Law*,—that the squares of the periodical times of the planets are as the cubes of the distances. Kepler's rapture on detecting it was unbounded. "What," says he, "I prophesied two-and-twenty years ago, as soon as I discovered the five solids among the heavenly orbits; what I firmly believed long before I had seen Ptolemy's Harmonics; what I had promised my friends in the title of this book, which I named before I was sure of my discovery; what, sixteen years ago, I urged as a thing to be sought; that for which I joined Tycho Brahe, for which I settled in Prague, for which I have devoted the best part of my life to astronomical contemplations;—at length I have brought to light, and have recognised its truth beyond my most sanguine expectations. It is now eighteen months since I got the first glimpse of light, three months since the dawn, very few days since the unveiled sun, most admirable to gaze on, burst out upon me. Nothing holds me: I will indulge in my sacred fury; I will triumph over mankind by the honest confession, that I have stolen the golden vases of the Egyptians to build up a tabernacle for my God, far from the confines of Egypt. If you forgive me, I rejoice: if you are angry, I can bear it; the die is cast, the book is written, to be read either now or by posterity,—I care not which. I may well wait a century for a reader, as God has waited six thousand years for an observer." In accordance with the notion he entertained respecting the "music of the spheres," he made Saturn and Jupiter take the bass, Mars the tenor, the Earth and Venus the counter, and Mercury the treble.

"The misery in which Kepler lived," says Sir David Brewster, in his 'Life of Newton,' "forms a painful contrast with the services which he performed for science. The pension on which he subsisted was always in arrears; and though the three emperors, whose reigns he adorned, directed their ministers to be more punctual in its payment, the disobedience of their commands was a source of continual vexation to Kepler. When he retired to Silesia, to spend the remainder of his days, his pecuniary difficulties became still more harassing. Necessity at length compelled him to apply personally for the arrears which were due; and he accordingly set out, in 1630, when nearly sixty years of age, for Ratisbon; but, in consequence of the great fatigue which so long a journey on horseback produced, he was seized with a fever, which put an end to his life."

Professor Whewell (in his interesting work on Astronomy and General Physics considered with reference to Natural Theology) expresses the opinion that Kepler, notwithstanding his constitutional oddities, was a man of strong and lively piety. His 'Commentaries on the Motions of Mars' he opens with the following passage: "I beseech my reader, that, not unmindful of the Divine goodness bestowed on man, he do with me praise and celebrate the wisdom and greatness of the Creator, which I open to him from a more inward explication of the form of the world, from a searching of causes, from a detection of the errors of vision; and that thus, not only in the firmness and stability of the earth, he perceive with gratitude the preservation of all living things in Nature as the gift of God, but also that in its motion, so recondite, so admirable, he acknowledge the wisdom of the Creator. But him who is too dull to receive this science, or too weak to believe the Copernican system without harm to his piety,—him, I say, I advise that, leaving the school of astronomy, and condemning, if he please, any doctrines of the philosophers, he follow his own path, and desist from this wandering through the universe; and, lifting up his natural eyes, with which he alone can see, pour himself out in his own heart, in praise of God the Creator; being certain that he gives no less worship to God than the astronomer, to whom God has given to see more clearly with his inward eye, and who, for what he has himself discovered, both can and will glorify God."

In a Life of Kepler, very recently published in his native country, founded on manuscripts of his which have lately been brought to light, there are given numerous other examples of a similar devotional spirit. Kepler thus concludes his Harmonics: "I give Thee thanks, Lord and Creator, that Thou has given me joy through Thy creation; for I have been ravished with the work of Thy hands. I have revealed unto mankind the glory of Thy works, as far as my limited spirit could conceive their infinitude. Should I have brought forward any thing that is unworthy of Thee, or should I have sought my own fame, be graciously pleased to forgive me."

As Galileo experienced the most bitter persecutions from the Church of Rome, so Kepler met with much violent opposition and calumny from the

Protestant clergy of his own country, particularly for adopting, in an almanac which, as astronomer royal, he annually published, the reformed calendar, as given by the Pope of Rome. His opinions respecting religious liberty, also, appear to have been greatly in advance of the times in which he lived. In answer to certain calumnies with which he was assailed, for his boldness in reasoning from the light of Nature, he uttered these memorable words: "The day will soon break, when pious simplicity will be ashamed of its blind superstition; when men will recognise truth in the book of Nature as well as in the Holy Scriptures, and rejoice in the two revelations."

LETTER XXV.

COMETS.

——"Fancy now no more
Wantons on fickle pinions through the skies,
But, fixed in aim, and conscious of her power,
Sublime from cause to cause exults to rise,
Creation's blended stores arranging as she flies."—*Beattie.*

NOTHING in astronomy is more truly admirable, than the knowledge which astronomers have acquired of the motions of comets, and the power they have gained of predicting their return. Indeed, every thing appertaining to this class of bodies is so wonderful, as to seem rather a tale of romance than a simple recital of facts. Comets are truly the knights-errant of astronomy. Appearing suddenly in the nocturnal sky, and often dragging after them a train of terrific aspect, they were, in the earlier ages of the world, and indeed until a recent period, considered as peculiarly ominous of the wrath of Heaven, and as harbingers of wars and famines, of the dethronement of monarchs, and the dissolution of empires.

Science has, it is true, disarmed them of their terrors, and demonstrated that they are under the guidance of the same Hand, that directs in their courses the other members of the solar system; but she has, at the same time, arrayed them in a garb of majesty peculiarly her own.

Although the ancients paid little attention to the ordinary phenomena of Nature, hardly deeming them worthy of a reason, yet, when a comet blazed forth, fear and astonishment conspired to make it an object of the most attentive observation. Hence the aspects of remarkable comets, that have appeared at various times, have been handed down to us, often with circumstantial minuteness, by the historians of different ages. The comet which appeared in the year 130, before the Christian era, at the birth of Mithridates, is said to have had a disk equal in magnitude to that of the sun. Ten years before this, one was seen, which, according to Justin, occupied a fourth part of the sky, that is, extended over forty-five degrees, and surpassed the sun in splendor. In the year 400, one was seen which resembled a sword in shape, and extended from the zenith to the horizon.

Such are some of the accounts of comets of past ages; but it is probable we must allow much for the exaggerations naturally accompanying the descriptions of objects in themselves so truly wonderful.

A comet, when perfectly formed, consists of three parts, the nucleus, the envelope, and the tail. The nucleus, or body of the comet, is generally distinguished by its forming a bright point in the centre of the head, conveying the idea of a solid, or at least of a very dense, portion of matter. Though it is usually exceedingly small, when compared with the other parts of the comet, and is sometimes wanting altogether, yet it occasionally subtends an angle capable of being measured by the telescope. The envelope (sometimes called the *coma*, from a Latin word signifying hair, in allusion to its hairy appearance) is a dense nebulous covering, which frequently renders the edge of the nucleus so indistinct, that it is extremely difficult to ascertain its diameter with any degree of precision. Many comets have no nucleus, but present only a nebulous mass, exceedingly attenuated on the confines, but gradually increasing in density towards the centre. Indeed, there is a regular gradation of comets, from such as are composed merely of a gaseous or vapory medium, to those which have a well-defined nucleus. In some instances on record, astronomers have detected with their telescopes small stars through the densest part of a comet. The tail is regarded as an expansion or prolongation of the coma; and presenting, as it sometimes does, a train of appalling magnitude, and of a pale, portentous light, it confers on this class of bodies their peculiar celebrity. These several parts are exhibited in Fig. 67, which represents the appearance of the comet of 1680. Fig. 68 also exhibits that of the comet of 1811.

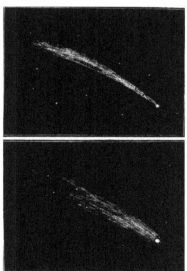

Figures 67, 68. COMETS OF 1680 AND 1811.

The *number* of comets belonging to the solar system, is probably very great. Many no doubt escape observation, by being above the horizon in the day-time. Seneca mentions, that during a total eclipse of the sun, which happened sixty years before the Christian era, a large and splendid comet suddenly made its appearance, being very near the sun. The leading particulars of at least one hundred and thirty have been computed, and arranged in a table, for future comparison. Of these, *six* are particularly remarkable; namely, the comets of 1680, 1770, and 1811; and those which bear the names of Halley, Biela, and Encke. The comet of 1680 was remarkable, not only for its astonishing size and splendor, and its near approach to the sun, but is celebrated for having submitted itself to the observations of Sir Isaac Newton, and for having enjoyed the signal honor of being the first comet whose elements were determined on the sure basis of mathematics. The comet of 1770 is memorable for the changes its orbit has undergone by the action of Jupiter, as I shall explain to you more particularly hereafter. The comet of 1811 was the most remarkable in its appearance of all that have been seen in the present century. It had scarcely any perceptible nucleus, but its train was very long and broad, as is represented in Fig. 68. Halley's comet (the same which reappeared in 1835) is distinguished as that whose return was first successfully predicted, and whose orbit is best determined; and Biela's and Encke's comets are well known for their short periods of revolution, which subject them frequently to the view of astronomers.

In *magnitude and brightness*, comets exhibit great diversity. History informs us of comets so bright, as to be distinctly visible in the day-time, even at noon, and in the brightest sunshine. Such was the comet seen at Rome a little before the assassination of Julius Cæsar. The comet of 1680 covered an arc of the heavens of ninety-seven degrees, and its length was estimated at one hundred and twenty-three millions of miles. That of 1811 had a nucleus of only four hundred and twenty-eight miles in diameter, but a tail one hundred and thirty-two millions of miles long. Had it been coiled around the earth like a serpent, it would have reached round more than five thousand times. Other comets are exceedingly small, the nucleus being in one case estimated at only twenty-five miles; and some, which are destitute of any perceptible nucleus, appear to the largest telescopes, even when nearest to us, only as a small speck of fog, or as a tuft of down. The majority of comets can be seen only by the aid of the telescope. Indeed, the same comet has very different aspects, at its different returns. Halley's comet, in 1305, was described by the historians of that age as the comet of terrific magnitude; (*cometa horrendæ magnitudinis*;) in 1456 its tail reached from the horizon to the zenith, and inspired such terror, that, by a decree of the Pope of Rome, public prayers were offered up at noonday in all the Catholic churches, to deprecate the wrath of heaven; while in 1682 its tail was only thirty degrees in length; and in 1759 it was visible only to the telescope until after it had passed its perihelion. At its recent return, in 1835, the greatest length of the tail was about twelve degrees. These changes in the appearance of the same comet are partly owing to the different positions of the earth with respect to them, being sometimes much nearer to them when they cross its track than at others; also, one spectator, so situated as to see the comet at a higher angle of elevation, or in a purer sky, than another, will see the train longer than it appears to another less favorably situated; but the extent of the changes are such as indicate also a real change in magnitude and brightness.

The *periods* of comets in their revolutions around the sun are equally various. Encke's comet, which has the shortest known period, completes its revolution in three and one third years; or, more accurately, in twelve hundred and eight days; while that of 1811 is estimated to have a period of thirty-three hundred and eighty three years.

The *distances* to which different comets recede from the sun are equally various. While Encke's comet performs its entire revolution within the orbit of Jupiter, Halley's comet recedes from the sun to twice the distance of Uranus; or nearly thirty-six hundred millions of miles. Some comets, indeed, are thought to go a much greater distance from the sun than this, while some are supposed to pass into curves which do not, like the ellipse, return into themselves; and in this case they never come back to the sun. (See Fig. 34, page 153.)

Comets shine *by reflecting the light of the sun.* In one or two instances, they have been thought to exhibit distinct *phases*, like the moon, although the nebulous matter with which the nucleus is surrounded would commonly prevent such phases from being distinctly visible, even when they would otherwise be apparent. Moreover, certain qualities of *polarized* light,—an affection by which a ray of light seems to have different properties on different sides,—enable opticians to decide whether the light of a given body is direct or reflected; and M. Arago, of Paris, by experiments of this kind on the light of the comet of 1819, ascertained it to be reflected light.

The tail of a comet usually increases very much as it approaches the sun; and it frequently does not reach its maximum until after the perihelion passage. In receding from the sun, the tail again contracts, and nearly or quite disappears before the body of the comet is entirely out of sight. The tail is frequently divided into two portions, the central parts, in the direction of the axis, being less bright than the marginal parts. In 1744 a comet appeared which had six tails spread out like a fan.

The tails of comets extend in a direct line from the sun, although more or less curved, like a long quill or feather, being convex on the side next to the direction in which they are moving,—a figure which may result from the less velocity of the portion most remote from the sun. Expansions of the envelope have also been at times observed on the side next the sun; but these seldom attain any considerable length.

The *quantity of matter* in comets is exceedingly small. Their tails consist of matter of such tenuity, that the smallest stars are visible through them. They can only be regarded as masses of thin vapor, susceptible of being penetrated through their whole substance by the sunbeams, and reflecting them alike from their interior parts and from their surfaces. It appears perhaps incredible, that so thin a substance should be visible by reflected light, and some astronomers have held that the matter of comets is self-luminous; but it requires but very little light to render an object visible in the night, and a light vapor may be visible when illuminated throughout an immense stratum, which could not be seen if spread over the face of the sky like a thin cloud. "The highest clouds that float in our atmosphere," says Sir John Herschel, "must be looked upon as dense and massive bodies, compared with the filmy and all but spiritual texture of a comet."

The small quantity of matter in comets is proved by the fact, that they have at times passed very near to some of the planets, without disturbing their motions in any appreciable degree. Thus the comet of 1770, in its way to the sun, got entangled among the satellites of Jupiter, and remained near them four months; yet it did not perceptibly change their motions. The same comet, also, came very near the earth; so that, had its quantity of matter been equal to that of the earth, it would, by its attraction, have caused the earth to revolve in an orbit so much larger than at present, as to have increased the length of the year two hours and forty-seven minutes. Yet it produced no sensible effect on the length of the year, and therefore its mass, as is shown by La Place, could not have exceeded 1/5000 of that of the earth, and might have been less than this to any extent. It may indeed be asked, what proof we have that comets have any matter, and are not mere reflections of light. The answer is, that, although they are not able by their own force of attraction to disturb the motions of the planets, yet they are themselves exceedingly disturbed by the action of the planets, and in exact conformity with the laws of universal gravitation. A delicate compass may be greatly agitated by the vicinity of a mass of iron, while the iron is not sensibly affected by the attraction of the needle.

By approaching very near to a large planet, a comet may have its orbit entirely changed. This fact is strikingly exemplified in the history of the comet of 1770. At its appearance in 1770, its orbit was found to be an ellipse, requiring for a complete revolution only five and a half years; and the wonder was, that it had not been seen before, since it was a very large and bright comet. Astronomers suspected that its path had been changed, and that it had been recently compelled to move in this short ellipse, by the disturbing force of Jupiter and his satellites. The French Institute, therefore, offered a high prize for the most complete investigation of the elements of this comet, taking into account any circumstances which could possibly have produced an alteration in its course. By tracing back the movements of this comet, for some years previous to 1770, it was found that, at the beginning of 1767, it had entered considerably within the sphere of Jupiter's attraction. Calculating the amount of this attraction from the known proximity of the two bodies, it was found what must have been its orbit previous to the time when it became subject to the disturbing action of Jupiter. It was therefore evident why, as long as it continued to circulate in an orbit so far from the centre of the system, it was never visible from the earth. In January, 1767, Jupiter and the comet happened to be very near to one another, and as both were moving in the same direction, and nearly in the same plane, they remained in the neighborhood of each other for several months, the planet being between the comet and the sun. The consequence was, that the comet's orbit was changed into a smaller ellipse, in which its revolution was accomplished in five and a half years. But as it approached the sun, in 1779, it happened again to fall in with Jupiter. It was in the month of June that the attraction of the planet began to have a sensible effect; and it was not until the month of October following, that they were finally separated.

At the time of their nearest approach, in August, Jupiter was distant from the comet only $1/_{491}$ of its distance from the sun, and exerted an attraction upon it two hundred and twenty-five times greater than that of the sun. By reason of this powerful attraction, Jupiter being further from the sun than the comet, the latter was drawn out into a new orbit, which even at it's perihelion came no nearer to the sun than the planet Ceres. In this third orbit, the comet requires about twenty years to accomplish it's revolution; and being at so great a distance from the earth, it is invisible, and will for ever remain so unless, in the course of ages, it may undergo new perturbations, and move again in some smaller orbit, as before.

With the foregoing leading facts respecting comets in view, I will now explain to you a few things equally remarkable respecting their *motions*.

The paths of the planets around the sun being nearly circular, we are able to see a planet in every part of it's orbit. But the case is very different with comets. For the greater part of their course, they are wholly out of sight, and come into view only while just in the neighborhood of the sun. This you will readily see must be the case, by inspecting the frontispiece, which represents the orbit of Biela's comet, in 1832. Sometimes, the orbit is so eccentric, that the place of the focus occupied by the sun appears almost at the extremity of the orbit. This was the case with the orbit of the comet of 1680. Indeed, this comet, at it's perihelion, came in fact nearer to the sun than the sixth part of the sun's diameter, being only one hundred and forty-six thousand miles from the surface of the sun, which, you will remark, is only a little more than half the distance of the moon from the earth; while, at it's aphelion, it was estimated to be thirteen thousand millions of miles from the sun,—more than eleven thousand millions of miles beyond the planet Uranus. Its *velocity*, when nearest the sun, exceeded a million of miles an hour. To describe such an orbit as was assigned to it by Sir Isaac Newton, would require five hundred and seventy-five years. During all this period, it was entirely out of view to the inhabitants of the earth, except the few months, while it was running down to the sun from such a distance as the orbit of Jupiter and back. The velocity of bodies moving in such eccentric orbits differs widely in different parts of their orbits. In the remotest parts it is so slow, that years would be required to pass over a space equal to that which it would run over in a single day, when near the sun.

The appearances of the same comet at different periods of it's return are so various, that we can never pronounce a given comet to be the same with one that has appeared before, from any peculiarities in it's physical aspect, as from it's color, magnitude, or shape; since, in all these respects, it is very different at different returns; but it is judged to be the same if it's *path* through the heavens, as traced among the stars, is the same.

The comet whose history is the most interesting, and which both of us have been privileged to see, is Halley's. Just before it's latest visit, in 1835, it's return was anticipated with so much expectation, not only by astronomers, but by all classes of the community, that a great and laudable eagerness universally prevailed, to learn the particulars of it's history. The best summary of these, which I met with, was given in the Edinburgh Review for April, 1835. I might content myself with barely referring you to that well-written article; but, as you may not have the work at hand, and would, moreover, probably not desire to read the whole article, I will abridge it for your perusal, interspersing some remarks of my own. I have desired to give you, in the course of these Letters, some specimen of the labors of astronomers, and shall probably never be able to find a better one.

It is believed that the first recorded appearance of Halley's comet was that which was supposed to signalize the birth of Mithridates, one hundred and thirty years before the birth of Christ. It is said to have appeared for twenty-four days; it's light is said to have surpassed that of the sun; it's magnitude to have extended over a fourth part of the firmament; and it is stated to have occupied, consequently, about four hours in rising and setting. In the year 323, a comet appeared in the sign Virgo. Another, according to the historians of the Lower Empire, appeared in the year 399, seventy-six years after the last, at an interval corresponding to that of Halley's comet. The interval between the birth of Mithridates and the year 323 was four hundred and fifty-three years, which would be equivalent to six periods of seventy-five and a half years. Thus it would seem, that in the interim there were five returns of this comet unobserved, or at least unrecorded. The appearance in the year 399 was attended with extraordinary circumstances. It was described in the old writers as a "comet of monstrous size and appalling aspect, it's tail seeming to reach down to the ground." The next recorded appearance of a comet agreeing with the ascertained period marks the taking of Rome, in the year 550,—an interval of one hundred and fifty-one years, or two periods of seventy-five and a half years having elapsed. One unrecorded return must, therefore, have taken place in the interim. The next appearance of a comet, coinciding with the assigned period, is three hundred and eighty years afterwards; namely, in the year 930,—five revolutions having been completed in the interval. The next appearance is recorded in the year 1005, after an interval of a single period of seventy-five years. Three revolutions would now seem to have passed unrecorded, when the comet again makes it's appearance in 1230. In this, as well as in former appearances, it is proper to state, that the sole test of identity of these cornets with that of Halley is the coincidence of the times, as near as historical records enable us to ascertain, with the epochs at which the comet of Halley might be expected to appear. That such evidence, however, is very imperfect, must be evident, if the frequency of cometary appearances be considered, and if it be remembered, that hitherto we find no recorded observations, which could enable us to trace, even with the rudest degree of approximation, the paths of those comets, the times of whose appearances raise a presumption of their identity with that of Halley. We now, however, descend to times in which more satisfactory evidence may be expected.

In the year 1305, a year in which the return of Halley's comet might have been expected, there is recorded a comet of remarkable character: "A comet of terrific dimensions made it's appearance about the time of the feast of the Passover, which was followed by a Great Plague." Had the terrific appearance of this body alone been recorded, this description might have passed without the charge of great exaggeration; but when we find the Great Plague connected with it as a consequence, it is impossible not to conclude, that the comet was seen by its historians through the magnifying medium of the calamity which followed it. Another appearance is recorded in the year 1380, unaccompanied by any other circumstance than its mere date. This, however, is in strict accordance with the ascertained period of Halley's comet.

We now arrive at the first appearance at which observations were taken, possessing sufficient accuracy to enable subsequent investigators to determine the path of the comet; and this is accordingly the first comet the identity of which with the comet of Halley can be said to be conclusively established. In the year 1456, a comet is stated to have appeared "of unheard of magnitude;" it was accompanied by a tail of extraordinary length, which extended over sixty degrees, (a third part of the heavens,) and continued to be seen during the whole month of June. The influence which was attributed to this appearance renders it probable, that in the record there is more or less of exaggeration. It was considered as the celestial indication of the rapid success of Mohammed the Second, who had taken Constantinople, and struck terror into the whole Christian world. Pope Calixtus the Second levelled the thunders of the Church against the enemies of his faith, terrestrial and celestial; and in the same Bull excommunicated the Turks and the comet; and, in order that the memory of this manifestation of his power should be for ever preserved, he ordained that the bells of all the churches should be rung at mid-day,—a custom which is preserved in those countries to our times.

The extraordinary length and brilliancy which was ascribed to the tail, upon this occasion, have led astronomers to investigate the circumstances under which its brightness and magnitude would be the greatest possible; and upon tracing back the motion of the comet to the year 1456, it has been found that it was then actually in the position, with respect to the earth and sun, most favorable to magnitude and splendor. So far, therefore, the result of astronomical calculation corroborates the records of history.

The next return took place in 1531. Pierre Appian, who first ascertained the fact that the tails of comets are usually turned from the sun, examined this comet with a view to verify his statement, and to ascertain the true direction of its tail. He made, accordingly, numerous observations upon its position, which, although rude, compared with the present standard of accuracy, were still sufficiently exact to enable Halley to identify this comet with that observed by himself.

The next return took place in 1607, when the comet was observed by Kepler. This astronomer first saw it on the evening of the twenty-sixth of September, when it had the appearance of a star of the first magnitude, and, to his vision, was without a tail; but the friends who accompanied him had better sight, and distinguished the tail. Before three o'clock the following morning the tail had become clearly visible, and had acquired great magnitude. Two days afterwards, the comet was observed by Longomontanus, a distinguished philosopher of the time. He describes its appearance, to the naked eye, to be like Jupiter, but of a paler and more obscured light; that its tail was of considerable length, of a paler light than that of the head, and more dense than the tails of ordinary comets.

The next appearance, and that which was observed by Halley himself, took place in 1682, a little before the publication of the '*Principia*.' In the interval between 1607 and 1682, practical astronomy had made great advances; instruments of observation had been brought to a state of comparative perfection; numerous observatories had been established, and the management of them had been confided to the most eminent men in Europe. In 1682, the scientific world was therefore prepared to examine the visitor of our system with a degree of care and accuracy before unknown.

In the year 1686, about four years afterwards, Newton published his '*Principia*,' in which he applied to the comet of 1680 the general principles of physical investigation first promulgated in that work. He explained the method of determining, by geometrical construction, the visible portion of the path of a body of this kind, and invited astronomers to apply these principles to the various recorded comets,—to discover whether some among them might not have appeared at different epochs, the future returns of which might consequently be predicted. Such was the effect of the force of analogy upon the mind of Newton, that, without awaiting the discovery of a periodic comet, he boldly assumed these bodies to be analogous to planets in their revolution round the sun.

Extraordinary as these conjectures must have appeared at the time, they were soon strictly realized. Halley, who was then a young man, but possessed one of the best minds in England, undertook the labor of examining the circumstances attending all the comets previously recorded, with a view to discover whether any, and which of them, appeared to follow the same path. Antecedently to the year 1700, four hundred and twenty-five of these bodies had been recorded in history; but those which had appeared before the fourteenth century had not been submitted to any observations by which their paths could be ascertained,—at least, not with a sufficient degree of precision, to afford any hope of identifying them with those of other comets. Subsequently to the year 1300, however, Halley found twenty-four comets on which observations had been made and recorded, with a degree of precision sufficient to enable him to calculate the actual paths which these bodies followed while they were visible. He examined, with the most elaborate care, the *courses* of each of these twenty-four bodies; he found the exact points at which each one of them crossed the ecliptic, or their *nodes*; also the angle which the direction of their motion made with that plane,—that is, the *inclination of their orbits*; he also calculated the nearest distance at which each of them approached the sun, or their *perihelion distance*; and the exact place of the body when at that nearest point,—that is, the *longitude of the perihelion*. These particulars are called the *elements* of a comet, because, when ascertained, they afford sufficient data for determining a comet's path. On comparing these paths, Halley found that one, which had appeared in 1661, followed nearly the same path as one which had appeared in 1532. Supposing, then, these to be two successive appearances of the same comet, it would follow, that its period would be one hundred and twenty-nine years, reckoning from 1661. Had this conjecture been well founded, the comet must have appeared about the year 1790. No comet, however, appeared at or near that time, following a similar path.

In his second conjecture, Halley was more fortunate, as indeed might be expected, since it was formed upon more conclusive grounds. He found that the paths of comets which had appeared in 1531 and 1607 were nearly identical, and that they were in fact the same as the path followed by the comet observed by himself in 1682. He suspected, therefore, that the appearances at these three epochs were produced by three successive returns of the same comet, and that, consequently, its period in its orbit must be about seventy-five and a half years. The probability of this conclusion is strikingly exhibited to the eye, by presenting the elements in a tabular form, from which it will at once be seen how nearly they correspond at these regular intervals.

Time.	Inclination of	Long. of the	Long. Per.	Per. Dist.	Course.
	the orbit.	node.			
1456	17°56'	48°30'	301°00'	0°58'	Retrograde.
1531	17 56	49 25	301 39	0 57	"
1607	17 02	50 21	302 16	0 58	"
1682	17 42	50 48	301 36	0 58	"

So little was the scientific world, at this time, prepared for such an announcement, that Halley himself only ventured at first to express his opinion in the form of conjecture; but, after some further investigation of the circumstances of the recorded comets, he found three which, at least in point of time, agreed with the period assigned to the comet of 1682. Collecting confidence from these circumstances, he announced his discovery as the result of observation and calculation combined, and entitled to as much confidence as any other consequence of an established physical law.

There were, nevertheless, two circumstances which might be supposed to offer some difficulty. First, the intervals between the supposed successive returns were not precisely equal; and, secondly, the inclination of the comet's path to the plane of the earth's orbit was not exactly the same in each case. Halley, however, with a degree of sagacity which, considering the state of knowledge at the time, cannot fail to excite unqualified admiration, observed, that it was natural to suppose that the same causes which disturbed the planetary motions must likewise act upon comets; and that their influence would be so much the more sensible upon these bodies, because of their great distances from the sun. Thus, as the attraction of Jupiter for Saturn was known to affect the velocity of the latter planet, sometimes retarding and sometimes accelerating it, according to their relative position, so as to affect its period to the extent of thirteen days, it might well be supposed, that the comet might suffer by a similar attraction an effect sufficiently great, to account for the inequality observed in the interval between its successive returns: and also for the variation to which the direction of its path upon the plane of the ecliptic was found to be subject. He observed, in fine, that, as in the interval between 1607 and 1682, the comet passed so near Jupiter that its velocity must have been augmented, and consequently its period shortened, by the action of that planet, this period, therefore, having been only seventy-five years, he inferred that the following period would probably be seventy-six years, or upwards; and consequently, that the comet ought not to be expected to appear until the end of 1758, or the beginning of 1759. It is impossible to imagine any quality of mind more enviable than that which, in the existing state of mathematical physics, could have led to such a prediction. The imperfect state of mathematical science rendered it impossible for Halley to offer to the world a demonstration of the event which he foretold. The theory of gravitation, which was in its infancy in the time of Halley's investigations, had grown to comparative maturity before the period at which his prediction could be fulfilled. The exigencies of that theory gave birth to new and more powerful instruments of mathematical inquiry: the differential and integral calculus, or the science of fluxions, as it is sometimes called,—a branch of the mathematics, expressed by algebraic symbols, but capable of a much higher reach, as an instrument of investigation, than either algebra or geometry,—was its first and greatest offspring. This branch of science was cultivated with an ardor and success by which it was enabled to answer all the demands of physics, and it contributed largely to the advancement of mechanical science itself, building upon the laws of motion a structure which has since been denominated 'Celestial Mechanics.' Newton's discoveries having obtained reception throughout the scientific world, his inquiries and his theories were followed up; and the consequences of the great principle of universal gravitation were rapidly developed. Since, according to this doctrine, *every body in nature attracts and is attracted by every other body*, it follows, that the comet was liable to be acted on by each of the planets, as well as by the sun,—a circumstance which rendered its movements much more difficult to follow, than would be the case were it subject merely to the projectile force and to the solar attraction. To estimate the time it would take for a ship to cross the Atlantic would be an easy task, were she subject to only one constant wind; but to estimate, beforehand, the exact influence which all other winds and the tides might have upon her passage, some accelerating and some retarding her course, would present a problem of the greatest difficulty. Clairaut, however, a celebrated French mathematician, undertook to estimate the effects that would be produced on Halley's comet by the attractions of all the planets. His aim was to investigate *general rules*, by which the computation could be made arithmetically, and hand them over to the practical calculator, to make the actual computations. Lalande, a practical astronomer, no less eminent in his own department, and who indeed first urged Clairaut to this inquiry, undertook the management of the astronomical and arithmetical part of the calculation. In this prodigious labor (for it was one of most appalling magnitude) he was assisted by the wife of an eminent watchmaker in Paris, named Lepaute, whose exertions on this occasion have deservedly registered her name in astronomical history.

It is difficult to convey to one who is not conversant with such investigations, an adequate notion of the labor which such an inquiry involved. The calculation of the influence of any one *planet* of the system upon any other is itself a problem of some complexity and difficulty; but still, one general computation, depending upon the calculation of the terms of a certain series, is sufficient for its solution. This comparative simplicity arises entirely from two circumstances which characterize the planetary orbits. These are, that, though they are ellipses, they differ very slightly from circles; and though the planets do not move in the plane of the ecliptic, yet none of them deviate considerably from that plane. But these characters do not belong to the orbits of comets, which, on the contrary, are highly eccentric, and make all possible angles with the ecliptic. The consequence of this is, that the calculation of the disturbances produced in the cometary orbits by the action of the planets must be conducted not like the planets, in one general calculation applicable to the whole orbits, but in a vast number of separate calculations; in which the orbit is considered, as it were, bit by bit, each bit requiring a calculation similar to the whole orbit of the planet. Now, when it is considered that the period of Halley's comet is about seventy-five years, and that every portion of its course, for two successive periods, was necessary to be calculated separately in this way, some notion may be formed of the labor encountered by Lalande and Madame Lepaute. "During six months," says Lalande, "we calculated from morning till night, sometimes even at meals; the consequence of which was, that I contracted an illness which changed my constitution for the remainder of my life. The assistance rendered by Madame Lepaute was such, that, without her, we never could have dared to undertake this enormous labor, in which it was necessary to calculate the distance of each of the two planets, Jupiter and Saturn, from the comet, and their attraction upon that body, separately, for every successive degree, and for one hundred and fifty years."

The attraction of a body is proportioned to its quantity of matter. Therefore, before the attraction exerted upon the comet by the several planets within whose influence it might fall, could be correctly estimated, it was necessary to know the mass of each planet; and though the planets had severally been weighed by methods supplied by Newton's 'Principia,' yet the estimate had not then attained the same measure of accuracy as it has now reached; nor was it certain that there was not (as it has since appeared that there actually was) one or more planets beyond Saturn, whose attractions might likewise influence the motions of the comet. Clairaut, making the best estimate he was able, under all these disadvantages, of the disturbing influence of the planets, fixed the return of the comet to the place of its nearest distance from the sun on the fourth of April, 1759.

In the successive appearances of the comet, subsequently to 1456, it was found to have gradually decreased in magnitude and splendor. While in 1456 it reached across one third part of the firmament, and spread terror over Europe, in 1607, its appearance, when observed by Kepler and Longomontanus, was that of a star of the first magnitude; and so trifling was its tail that, Kepler himself, when he first saw it, doubted whether it had any. In 1682, it excited little attention, except among astronomers. Supposing this decrease of magnitude and brilliancy to be progressive, Lalande entertained serious apprehensions that on its expected return it might be so inconsiderable, as to escape the observation even of astronomers; and thus, that this splendid example of the power of science, and unanswerable proof of the principle of gravitation, would be lost to the world.

It is not uninteresting to observe the misgivings of this distinguished astronomer with respect to the appearance of the body, mixed up with his unshaken faith in the result of the astronomical inquiry. "We cannot doubt," says he, "that it will return; and even if astronomers cannot see it, they will not therefore be the less convinced of its presence. They know that the faintness of its light, its great distance, and perhaps even bad weather, may keep it from our view. But the world will find it difficult to believe us; they will place this discovery, which has done so much honor to modern philosophy, among the number of chance predictions. We shall see discussions spring up again in colleges, contempt among the ignorant, terror among the people; and seventy-six years will roll away, before there will be another opportunity of removing all doubt."

Fortunately for science, the arrival of the expected visitor did not take place under such untoward circumstances. As the commencement of the year 1759 approached, "astronomers," says Voltaire, "hardly went to bed at all." The honor, however, of the first glimpse of the stranger was not reserved for the possessors of scientific rank, nor for the members of academies or universities. On the night of Christmas-day, 1758, George Palitzch, of Politz, near Dresden,—"a peasant," says Sir John Herchel, "by station, an astronomer by nature," first saw the comet.

An astronomer of Leipzic found it soon after; but, with the mean jealousy of a miser, he concealed his treasure, while his contemporaries throughout Europe were vainly directing their anxious search after it to other quarters of the heavens. At this time, Delisle, a French astronomer, and his assistant, Messier, who, from his unweared assiduity in the pursuit of comets, was called the *Comet-Hunter*, had been constantly engaged, for eighteen months, in watching for the return of Halley's comet. Messier passed his life in search of comets. It is related of him, that when he was in expectation of discovering a comet, his wife was taken ill and died. While attending on her, being withdrawn from his observatory, another astronomer anticipated him in the discovery. Messier was in despair. A friend, visiting him, began to offer some consolation for the recent affliction he had suffered. Messier, thinking only of the comet, exclaimed, "I had discovered twelve: alas, that I should be robbed of the thirteenth by Montague!"—and his eyes filled with tears. Then, remembering that it was necessary to mourn for his wife, whose remains were still in the house, he exclaimed, "Ah! this poor woman!" (*ah! cette pauvre femme,*) and again wept for his comet. We can easily imagine how eagerly such an enthusiast would watch for Halley's comet; and we could almost wish that it had been his good fortune to be the first to announce its arrival: but, being misled by a chart which directed his attention to the wrong part of the firmament, a whole month elapsed after its discovery by Palitzch, before he enjoyed the delightful spectacle.

The comet arrived at its perihelion on the thirteenth of March, only twenty-three days from the time assigned by Clairaut. It appeared very round, with a brilliant nucleus, well distinguished from the surrounding nebulosity. It had, however, no appearance of a tail. It became lost in the sun, as it approached its perihelion, and emerged again, on the other side of the sun, on the first of April. Its exhibiting an appearance, so inferior to what it presented on some of its previous returns, is partly accounted for by its being seen by the European astronomers under peculiarly disadvantageous circumstances, being almost always within the twilight, and in the most unfavorable situations. In the southern hemisphere, however, the circumstances for observing it were more favorable, and there it exhibited a tail varying from ten to forty-seven degrees in length.

In my next Letter I will give you some particulars respecting the late return of Halley's comet.

LETTER XXVI.

COMETS, CONTINUED.

"Incensed with indignation, Satan stood
Unterrified, and like a comet burned,
That fires the length of Ophiucus huge
In the Arctic sky, and from his horrid train
Shakes pestilence and war."—*Milton.*

AMONG other great results which have marked the history of Halley's comet, it has itself been a criterion of the existing state of the mathematical and astronomical sciences. We have just seen how far the knowledge of the great laws of physical astronomy, and of the higher mathematics, enabled the astronomers of 1682 and 1759, respectively, to deal with this wonderful body; and let us now see what higher advantages were possessed by the astronomers of 1835. During this last interval of seventy-six years, the science of mathematics, in its most profound and refined branches, has made prodigious advances, more especially in its application to the laws of the celestial motions, as exemplified in the 'Mecanique Celeste' of La Place. The methods of investigation have acquired greater simplicity, and have likewise become more general and comprehensive; and mechanical science, in the largest sense of that term, now embraces in its formularies the most complicated motions, and the most minute effects of the mutual influences of the various members of our system. You will probably find it difficult to comprehend, how such hidden facts can be disclosed by formularies, consisting of a's and b's, and x's and y's, and other algebraic symbols; nor will it be easy to give you a clear idea of this subject, without a more extensive acquaintance than you have formed with algebraic investigations; but you can easily understand that even an equation expressed in numbers may be so changed in its form, by adding, subtracting, multiplying and dividing, as to express some new truth at every transformation. Some idea of this may be formed by the simplest example. Take the following: $3+4=7$. This equation expresses the fact, that three added to four is equal to seven. By multiplying all the terms by 2, we obtain a new equation, in which $6+8=14$. This expresses a new truth; and by varying the form, by similar operations, an indefinite number of separate truths may be elicited from the simple fundamental expression. I will add another illustration, which involves a little more algebra, but not, I think, more than you can understand; or, if it does, you will please pass over it to the next paragraph. According to a rule of arithmetical progression, *the sum of all the terms is equal to half the sum of the extremes multiplied into the number of terms.* Calling the sum of the terms s, the first term a, the last b, and the number of terms n, and we have $(1/2)n(a+b)=s$; or $n(a+b)=2s$; or $a+b=\frac{2s}{n}$; or $a=(\frac{2s}{n})-b$; or $b=(\frac{2s}{n})-a$. These are only a few of the changes which may be made in the original expression, still preserving the equality between the quantities on the left hand and those on the right; yet each of these transformations expresses a new truth, indicating distinct and (as might be the case) before unknown relations between the several quantities of which the whole expression is composed. The last, for example, shows us that the last term in an arithmetical series is always equal to twice the sum of the whole series divided by the number of terms and diminished by the first term. In analytical formularies, as expressions of this kind are called, the value of a single unknown quantity is sometimes given in a very complicated expression, consisting of known quantities; but before we can ascertain their united value, we must reduce them, by actually performing all the additions, subtractions, multiplications, divisions, raising to powers, and extracting roots, which are denoted by the symbols. This makes the actual calculations derived from such formularies immensely laborious. We have already had an instance of this in the calculations made by Lalande and Madame Lepaute, from formularies furnished by Clairaut.

The analytical formularies, contained in such works as La Place's 'Mecanique Celeste,' exhibit to the eye of the mathematician a record of all the evolutions of the bodies of the solar system in ages past, and of all the changes they must undergo in ages to come. Such has been the result of the combination of transcendent mathematical genius and unexampled labor and perseverance, for the last century. The learned societies established in various centres of civilization have more especially directed their attention to the advancement of physical astronomy, and have stimulated the spirit of inquiry by a succession of prizes, offered for the solutions of problems arising out of the difficulties which were progressively developed by the advancement of astronomical knowledge. Among these questions, the determination of the return of comets, and the disturbances which they experience in their course, by the action of the planets near which they happen to pass, hold a prominent place. In 1826, the French Institute offered a prize for the determination of the exact time of the return of Halley's comet to its perihelion in 1835. M. Pontecoulant aspired to the honor. "After calculations," says he, "of which those alone who have engaged in such researches can estimate the extent and appreciate the fastidious monotony, I arrived at a result which satisfied all the conditions proposed by the Institute. I determined the perturbations of Halley's comet, by taking into account the simultaneous actions of Jupiter, Saturn, Uranus, and the Earth, and I then fixed its return to its perihelion for the seventh of November." Subsequently to this, however, M. Pontecoulant made some further researches, which led him to correct the former result; and he afterwards altered the time to November fourteenth. It actually came to its perihelion on the sixteenth, within two days of the time assigned.

Nothing can convince us more fully of the complete mastery which astronomers have at last acquired over these erratic bodies, than to read in the Edinburgh Review for April, 1835, the paragraph containing the final results of all the labors and anticipations of astronomers, matured as they were, in readiness for the approaching visitant, and then to compare the prediction with the event, as we saw it fulfilled a few months afterwards. The paragraph was as follows: "On the whole, it may be considered as tolerably certain, that the comet will become visible in every part of Europe about the latter end of August, or beginning of September, next. It will most probably be distinguishable by the naked eye, like a star of the first magnitude, but with a duller light than that of a planet, and surrounded with a pale nebulosity, which will slightly impair its splendor. On the night of the seventh of October, the comet will approach the well-known constellation of the Great Bear; and between that and the eleventh, it will pass directly through the seven conspicuous stars of that constellation, (the Dipper.) Towards the end of November, the comet will plunge among the rays of the sun, and disappear, and will not issue from them, on the other side, until the end of December."

Let us now see how far the actual appearances corresponded to these predictions. The comet was first discovered from the observatory at Rome, on the morning of the fifth of August; by Professor Struve, at Dorpat, on the twentieth; in England and France, on the twenty-third; and at Yale College, by Professor Loomis and myself, on the thirty-first. On the morning of that day, between two and three o'clock, in obedience to the directions which the great minds that had marked out its path among the stars had prescribed, we directed Clarke's telescope (a noble instrument, belonging to Yale College) towards the northeastern quarter of the heavens, and lo! there was the wanderer so long foretold,—a dim speck of fog on the confines of creation. It came on slowly, from night to night, increasing constantly in magnitude and brightness, but did not become distinctly visible to the naked eye until the twenty-second of September. For a month, therefore, astronomers enjoyed this interesting spectacle before it exhibited itself to the world at large. From this time it moved rapidly along the northern sky, until, about the tenth of October, it traversed the constellation of the Great Bear, passing a little above, instead of "through" the seven conspicuous stars constituting the Dipper. At this time it had a lengthened train, and became, as you doubtless remember, an object of universal interest. Early in November, the comet ran down to the sun, and was lost in his beams; but on the morning of December thirty-first, I again obtained, through Clarke's telescope, a distinct view of it on the other side of the sun, a moment before the morning dawn.

This return of Halley's comet was an astronomical event of transcendent importance. It was the chronicler of ages, and carried us, by a few steps, up to the origin of time. If a gallant ship, which has sailed round the globe, and commanded successively the admiration of many great cities, diverse in language and customs, is invested with a peculiar interest, what interest must attach to one that has made the circuit of the solar system, and fixed the gaze of successive worlds! So intimate, moreover, is the bond which binds together all truths in one indissoluble chain, that the establishment of one great truth often confirms a multitude of others, equally important. Thus the return of Halley's comet, in exact conformity with the predictions of astronomers, established the truth of all those principles by which those predictions were made. It afforded most triumphant proof of the doctrine of universal gravitation, and of course of the received laws of physical astronomy; it inspired new confidence in the power and accuracy of that instrument (the calculus) by means of which its elements had been investigated; and it proved that the different planets, which exerted upon it severally a disturbing force proportioned to their quantity of matter, had been correctly weighed, as in a balance.

I must now leave this wonderful body to pursue its sublime march far beyond the confines of Uranus, (a distance it has long since reached,) and take a hasty notice of two other comets, whose periodic returns have also been ascertained; namely, those of Biela and Encke.

Biela's comet has a period of six years and three quarters. It has its perihelion near the orbit of the earth, and its aphelion a little beyond that of Jupiter. Its orbit, therefore, is far less eccentric than that of Halley's comet; (see Frontispiece;) it neither approaches so near the sun, nor departs so far from it, as most other known comets: some, indeed, never come nearer to the sun than the orbit of Jupiter, while they recede to an incomprehensible distance beyond the remotest planet. We might even imagine that they would get beyond the limits of the sun's attraction; nor is this impossible, although, according to La Place, the solar attraction is sensible throughout a sphere whose radius is a hundred millions of times greater than the distance of the earth from the sun, or nearly ten thousand billions of miles.

Some months before the expected return of Biela's comet, in 1832, it was announced by astronomers, who had calculated its path, that it would cross the plane of the earth's orbit very near to the earth's path, so that, should the earth happen at the time to be at that point of her revolution, a collision might take place. This announcement excited so much alarm among the ignorant classes in France, that it was deemed expedient by the French academy, that one of their number should prepare and publish an article on the subject, with the express view of allaying popular apprehension. This task was executed by M. Arago. He admitted that the earth would in fact pass so near the point where the comet crossed the plane of its orbit, that, should they chance to meet there, the earth would be enveloped in the nebulous atmosphere of the comet. He, however, showed that the earth would not be near that point at the same time with the comet, but fifty millions of miles from it.

The comet came at the appointed time, but was so exceedingly faint and small, that it was visible only to the largest telescopes. In one respect, its diminutive size and feeble light enhanced the interest with which it was contemplated; for it was a sublime spectacle to see a body, which, as projected on the celestial vault, even when magnified a thousand times, seemed but a dim speck of fog, still pursuing its way, in obedience to the laws of universal gravitation, with the same regularity as Jupiter and Saturn. We are apt to imagine that a body, consisting of such light materials that it can be compared only to the thinnest fog, would be dissipated and lost in the boundless regions of space; but so far is this from the truth, that, when subjected to the action of the same forces of projection and solar attraction, it will move through the void regions of space, and will describe its own orbit about the sun with the same unerring certainty, as the densest bodies of the system.

Encke's comet, by its frequent returns, (once in three and a third years,) affords peculiar facilities for ascertaining the laws of its revolution; and it has kept the appointments made for it with great exactness. On its return in 1839, it exhibited to the telescope a globular mass of nebulous matter, resembling fog, and moved towards its perihelion with great rapidity. It makes its entire excursions within the orbit of Jupiter.

But what has made Encke's comet particularly famous, is its having first revealed to us the existence of a *resisting medium* in the planetary spaces. It has long been a question, whether the earth and planets revolve in a perfect void, or whether a fluid of extreme rarity may not be diffused through space. A perfect vacuum was deemed most probable, because no such effects on the motions of the planets could be detected as indicated that they encountered a resisting medium. But a feather, or a lock of cotton, propelled with great velocity, might render obvious the resistance of a medium which would not be perceptible in the motions of a cannon ball. Accordingly, Encke's comet is thought to have plainly suffered a retardation from encountering a resisting medium in the planetary regions.

The effect of this resistance, from the first discovery of the comet to the present time, has been to diminish the time of its revolution about two days. Such a resistance, by destroying a part of the projectile force, would cause the comet to approach nearer to the sun, and thus to have its periodic time shortened. The ultimate effect of this cause will be to bring the comet nearer to the sun, at every revolution, until it finally falls into that luminary, although many thousand years will be required to produce this catastrophe. It is conceivable, indeed, that the effects of such a resistance may be counteracted by the attraction of one or more of the planets, near which it may pass in its successive returns to the sun. Still, it is not probable that this cause will exactly counterbalance the other; so that, if there is such an elastic medium diffused through the planetary regions, it must follow that, in the lapse of ages, every comet will fall into the sun. Newton conjectured that this would be the case, although he did not found his opinion upon the existence of such a resisting medium as is now detected. To such an opinion he adhered to the end of life. At the age of eighty-three, in a conversation with his nephew, he expressed himself thus: "I cannot say when the comet of 1680 will fall into the sun; possibly after five or six revolutions; but whenever that time shall arrive, the heat of the sun will be raised by it to such a point, that our globe will be burned, and all the animals upon it will perish."

Of the *physical nature* of comets little is understood. The greater part of them are evidently mere masses of vapor, since they permit very small stars to be seen through them. In September, 1832, Sir John Herschel, when observing Biela's comet, saw that body pass directly between his eye and a small cluster of minute telescopic stars of the sixteenth or seventeenth magnitude. This little constellation occupied a space in the heavens, the breadth of which was not the twentieth part of that of the moon; yet the whole of the cluster was distinctly visible through the comet. "A more striking proof," says Sir John Herschel, "could not have been afforded, of the extreme transparency of the matter of which this comet consists. The most trifling fog would have entirely effaced this group of stars, yet they continued visible through a thickness of the comet which, calculating on its distance and apparent diameter, must have exceeded fifty thousand miles, at least towards its central parts." From this and similar observations, it is inferred, that the nebulous matter of comets is vastly more rare than that of the air we breathe, and hence, that, were more or less of it to be mingled with the earth's atmosphere, it would not be perceived, although it might possibly render the air unwholesome for respiration. M. Arago, however, is of the opinion, that some comets, at least, have a solid nucleus. It is difficult, on any other supposition, to account for the strong light which some of them have exhibited,—a light sufficiently intense to render them visible in the day-time, during the presence of the sun. The intense heat to which comets are subject, in approaching so near the sun as some of them do, is alleged as a sufficient reason for the great expansion of the thin vapory atmospheres which form their tails; and the inconceivable cold to which they are subject, in receding to such a distance from the sun, is supposed to account for the condensation of the same matter until it returns to its original dimensions. Thus the great comet of 1680, at its perihelion, approached within one hundred and forty-six thousand miles of the surface of the sun, a distance of only one sixth part of the sun's diameter. The heat which it must have received was estimated to be equal to twenty-eight thousand times that which the earth receives in the same time, and two thousand times hotter than red-hot iron. This temperature would be sufficient to volatilize the most obdurate substances, and to expand the vapor to vast dimensions; and the opposite effects of the extreme cold to which it would be subject in the regions remote from the sun would be adequate to condense it into its former volume. This explanation, however, does not account for the direction of the tail, extending, as it usually does, only in a line opposite to the sun. Some writers, therefore, suppose that the nebulous matter of the comet, after being expanded to such a volume that the particles are no longer attracted to the nucleus, unless by the slightest conceivable force, are carried off in a direction from the sun, by the impulse of the solar rays themselves. But to assign such a power to the sun's rays, while they have never been proved to have any momentum, is unphilosophical; and we are compelled to place the phenomena of comets' tails among the points of astronomy yet to be explained.

Since comets which approach very near the sun, like the comet of 1680, cross the orbits of all the planets, the possibility that one of them may strike the earth has frequently been suggested. Still it may quiet our apprehensions on this subject, to reflect on the vast amplitude of the planetary spaces, in which these bodies are not crowded together, as we see them erroneously represented in orreries and diagrams, but are sparsely scattered at immense distances from each other. They are like insects flying, singly, in the expanse of heaven. If a comet's tail lay with its axis in the plane of the ecliptic when it was near the sun, we can imagine that the tail might sweep over the earth; but the tail may be situated at any angle with the ecliptic, as well as in the same plane with it, and the chances that it will not be in the same plane are almost infinite. It is also extremely improbable that a comet will cross the plane of the ecliptic precisely at the earth's path in that plane, since it may as probably cross it at any other point nearer or more remote from the sun. A French writer of some eminence (Du Sejour) has discussed this subject with ability, and arrived at the following conclusions: That of all the comets whose paths had been ascertained, none *could pass* nearer to the earth than about twice the moon's distance; and that none ever *did pass* nearer to the earth than nine times the moon's distance. The comet of 1770, already mentioned, which became entangled among the satellites of Jupiter, came within this limit. Some have taken alarm at the idea that a comet, by approaching very near to the earth, might raise so high a *tide*, as to endanger the safety of maritime countries especially: but this writer shows, that the comet could not possibly remain more than two hours so near the earth as a fourth part of the moon's distance; and it could not remain even so long, unless it passed the earth under very peculiar circumstances. For example, if its orbit were nearly perpendicular to that of the earth, it could not remain more than half an hour in such a position. Under such circumstances, the production of a tide would be impossible. Eleven hours, at least, would be necessary to enable a comet to produce an effect on the waters of the earth, from which the injurious effects so much dreaded would follow. The final conclusion at which he arrives is, that although, in strict geometrical rigor, it is not physically impossible that a comet should encounter the earth, yet the probability of such an event is absolutely nothing.

M. Arago, also, has investigated the probability of such a collision on the mathematical doctrine of chances, and remarks as follows: "Suppose, now, a comet, of which we know nothing but that, at its perihelion, it will be nearer the sun than we are, and that its diameter is equal to one fourth that of the earth; the doctrine of chances shows that, out of two hundred and eighty-one millions of cases, there is but one against us; but one, in which the two bodies could meet."

La Place has assigned the consequences that would result from a direct collision between the earth and a comet. "It is easy," says he, "to represent the effects of the shock produced by the earth's encountering a comet. The axis and the motion of rotation changed; the waters abandoning their former position to precipitate themselves towards the new equator; a great part of men and animals whelmed in a universal deluge, or destroyed by the violent shock imparted to the terrestrial globe; entire species annihilated; all the monuments of human industry overthrown;—such are the disasters which the shock of a comet would necessarily produce." La Place, nevertheless, expresses a decided opinion that the orbits of the planets have never yet been disturbed by the influence of comets. Comets, moreover, have been, and are still to some degree, supposed to exercise much influence in the affairs of this world, affecting the weather, the crops, the public health, and a great variety of atmospheric commotions. Even Halley, finding that his comet must have been near the earth at the time of the Deluge, suggested the possibility that the comet caused that event,—an idea which was taken up by Whiston, and formed into a regular theory. In Gregory's Astronomy, an able work, published at Oxford in 1702, the author remarks, that among all nations and in all ages, it has been observed, that the appearance of a comet has always been followed by great calamities; and he adds, "it does not become philosophers lightly to set down these things as fables." Among the various things ascribed to comets by a late English writer, are hot and cold seasons, tempests, hurricanes, violent hail-storms, great falls of snow, heavy rains, inundations, droughts, famines, thick fogs, flies, grasshoppers, plague, dysentery, contagious diseases among animals, sickness among cats, volcanic eruptions, and meteors, or shooting stars. These notions are too ridiculous to require a distinct refutation; and I will only add, that we have no evidence that comets have hitherto ever exercised the least influence upon the affairs of this world; and we still remain in darkness, with respect to their physical nature, and the purposes for which they were created.

LETTER XXVII.

METEORIC SHOWERS.

"Oft shalt thou see, ere brooding storms arise,
Star after star glide headlong down the skies,
And, where they shot, long trails of lingering light
Sweep far behind, and gild the shades of night."—*Virgil*.

FEW subjects of astronomy have excited a more general interest, for several years past, than those extraordinary exhibitions of shooting stars, which have acquired the name of meteoric showers. My reason for introducing the subject to your notice, in this place, is, that these small bodies are, as I believe, derived from nebulous or cometary bodies, which belong to the solar system, and which, therefore, ought to be considered, before we take our leave of this department of creation, and naturally come next in order to comets.

The attention of astronomers was particularly directed to this subject by the extraordinary shower of meteors which occurred on the morning of the thirteenth of November, 1833. I had the good fortune to witness these grand celestial fire-works, and felt a strong desire that a phenomenon, which, as it afterwards appeared, was confined chiefly to North America, should here command that diligent inquiry into its causes, which so sublime a spectacle might justly claim.

As I think you were not so happy as to witness this magnificent display, I will endeavor to give you some faint idea of it, as it appeared to me a little before daybreak. Imagine a constant succession of fire-balls, resembling sky-rockets, radiating in all directions from a point in the heavens a few degrees southeast of the zenith, and following the arch of the sky towards the horizon. They commenced their progress at different distances from the radiating point; but their directions were uniformly such, that the lines they described, if produced upwards, would all have met in the same part of the heavens. Around this point, or imaginary radiant, was a circular space of several degrees, within which no meteors were observed. The balls, as they travelled down the vault, usually left after them a vivid streak of light; and, just before they disappeared, exploded, or suddenly resolved themselves into smoke. No report of any kind was observed, although we listened attentively.

Beside the foregoing distinct concretions, or individual bodies, the atmosphere exhibited *phosphoric lines*, following in the train of minute points, that shot off in the greatest abundance in a northwesterly direction. These did not so fully copy the figure of the sky, but moved in paths more nearly rectilinear, and appeared to be much nearer the spectator than the fire-balls. The light of their trains was also of a paler hue, not unlike that produced by writing with a stick of phosphorus on the walls of a dark room. The number of these luminous trains increased and diminished alternately, now and then crossing the field of view, like snow drifted before the wind, although, in fact, their course was towards the wind.

From these two varieties, we were presented with meteors of various sizes and degrees of splendor: some were mere points, while others were larger and brighter than Jupiter or Venus; and one, seen by a credible witness, at an earlier hour, was judged to be nearly as large as the moon. The flashes of light, although less intense than lightning, were so bright, as to awaken people in their beds. One ball that shot off in the northwest direction, and exploded a little northward of the star Capella, left, just behind the place of explosion, a phosphorescent train of peculiar beauty. This train was at first nearly straight, but it shortly began to contract in length, to dilate in breadth, and to assume the figure of a serpent drawing itself up, until it appeared like a small luminous cloud of vapor. This cloud was borne eastward, (by the wind, as was supposed, which was blowing gently in that direction,) opposite to the direction in which the meteor itself had moved, remaining in sight several minutes. The point from which the meteors seemed to radiate kept a fixed position among the stars, being constantly near a star in Leo, called Gamma Leonis.

Such is a brief description of this grand and beautiful display, as I saw it at New Haven. The newspapers shortly brought us intelligence of similar appearances in all parts of the United States, and many minute descriptions were published by various observers; from which it appeared, that the exhibition had been marked by very nearly the same characteristics wherever it had been seen. Probably no celestial phenomenon has ever occurred in this country, since its first settlement, which was viewed with so much admiration and delight by one class of spectators, or with so much astonishment and fear by another class. It strikingly evinced the progress of knowledge and civilization, that the latter class was comparatively so small, although it afforded some few examples of the dismay with which, in barbarous ages of the world, such spectacles as this were wont to be regarded. One or two instances were reported, of persons who died with terror; many others thought the last great day had come; and the untutored black population of the South gave expression to their fears in cries and shrieks.

After collecting and collating the accounts given in all the periodicals of the country, and also in numerous letters addressed either to my scientific friends or to myself, the following appeared to be the *leading facts* attending the phenomenon. The shower pervaded nearly the whole of North America, having appeared in nearly equal splendor from the British possessions on the north to the West-India Islands and Mexico on the south, and from sixty-one degrees of longitude east of the American coast, quite to the Pacific Ocean on the west. Throughout this immense region, the duration was nearly the same. The meteors began to attract attention by their unusual frequency and brilliancy, from *nine to twelve* o'clock in the evening; were most striking in their appearance from *two to five;* arrived at their maximum, in many places, about *four* o'clock; and continued until rendered invisible by the light of day. The meteors moved either in right lines, or in such apparent curves, as, upon optical principles, can be resolved into right lines. Their general tendency was towards the northwest, although, by the effect of perspective, they appeared to move in various directions.

Such were the leading phenomena of the great meteoric shower of November 13, 1833. For a fuller detail of the facts, as well as of the reasonings that were built on them, I must beg leave to refer you to some papers of mine in the twenty-fifth and twenty-sixth volumes of the American Journal of Science.

Soon after this wonderful occurrence, it was ascertained that a similar meteoric shower had appeared in 1799, and, what was remarkable, almost at exactly the same time of year, namely, on the morning of the twelfth of November; and we were again surprised as well as delighted, at receiving successive accounts from different parts of the world of the phenomenon, as having occurred on the morning of the same thirteenth of November, in 1830, 1831, and 1832. Hence this was evidently an event independent of the casual changes of the atmosphere; for, having a periodical return, it was undoubtedly to be referred to astronomical causes, and its recurrence, at a certain definite period of the year, plainly indicated *some* relation to the revolution of the earth around the sun. It remained, however, to develope the nature of this relation, by investigating, if possible, the origin of the meteors. The views to which I was led on this subject suggested the probability that the same phenomenon would recur on the corresponding seasons of the year, for at least several years afterwards; and such proved to be the fact, although the appearances, at every succeeding return, were less and less striking, until 1839, when, so far as I have heard, they ceased altogether.

Mean-while, two other distinct periods of meteoric showers have, as already intimated, been determined; namely, about the ninth of August, and seventh of December. The facts relative to the history of these periods have been collected with great industry by Mr. Edward C. Herrick; and several of the most ingenious and most useful conclusions, respecting the laws that regulate these singular exhibitions, have been deduced by Professor Twining. Several of the most distinguished astronomers of the Old World, also, have engaged in these investigations with great zeal, as Messrs. Arago and Biot, of Paris; Doctor Olbers, of Bremen; M. Wartmann, of Geneva; and M. Quetelet, of Brussels.

But you will be desirous to learn what are the *conclusions* which have been drawn respecting these new and extraordinary phenomena of the heavens. As the inferences to which I was led, as explained in the twenty-sixth volume of the 'American Journal of Science,' have, at least in their most important points, been sanctioned by astronomers of the highest respectability, I will venture to give you a brief abstract of them, with such modifications as the progress of investigation since that period has rendered necessary.

The principal questions involved in the inquiry were the following:— Was the *origin* of the meteors within the atmosphere, or beyond it? What was the *height* of the place above the surface of the earth? By what *force* were the meteors drawn or impelled towards the earth? In what *directions* did they move? With what *velocity*? What was the cause of their *light* and *heat*? Of what *size* were the larger varieties? At what height above the earth did they *disappear*? What was the nature of the *luminous trains* which sometimes remained behind? What *sort of bodies* were the meteors themselves; of what *kind of matter* constituted; and in what manner did they exist *before they fell to the earth*? Finally, what *relations* did the source from which they emanated sustain to our earth?

In the first place, *the meteors had their origin beyond the limits of our atmosphere.* We know whether a given appearance in the sky is within the atmosphere or beyond it, by this circumstance: all bodies near the earth, including the atmosphere itself, have a common motion with the earth around its axis from west to east. When we see a celestial object moving regularly from west to east, at the same rate as the earth moves, leaving the stars behind, we know it is near the earth, and partakes, in common with the atmosphere, of its diurnal rotation: but when the earth leaves the object behind; or, in other words, when the object moves westward along with the stars, then we know that it is so distant as not to participate in the diurnal revolution of the earth, and of course to be beyond the atmosphere. The source from which the meteors emanated thus kept pace with the stars, and hence was beyond the atmosphere.

In the second place, *the height of the place whence the meteors proceeded was very great, but it has not yet been accurately determined.* Regarding the body whence the meteors emanated after the similitude of a cloud, it seemed possible to obtain its height in the same manner as we measure the height of a cloud, or indeed the height of the moon. Although we could not see the body itself, yet the part of the heavens whence the meteors came would indicate its position. This point we called the *radiant*; and the question was, whether the radiant was projected by distant observers on different parts of the sky; that is, whether it had any *parallax*. I took much pains to ascertain the truth of this matter, by corresponding with various observers in different parts of the United States, who had accurately noted the position of the radiant among the fixed stars, and supposed I had obtained such materials as would enable us to determine the parallax, at least approximately; although such discordances existed in the evidence as reasonably to create some distrust of its validity. Putting together, however, the best materials I could obtain, I made the height of the radiant above the surface of the earth *twenty-two hundred and thirty-eight miles.* When, however, I afterwards obtained, as I supposed, some insight into the celestial origin of the meteors, I at once saw that the meteoric body must be much further off than this distance; and my present impression is, that we have not the means of determining what its height really is. We may safely place it at many thousand miles.

In the third place, with respect to the *force* by which the meteors were *drawn* or impelled towards the earth, my first impression was, that they fell merely by the force of *gravity*; but the velocity which, on careful investigation by Professor Twining and others, has been ascribed to them, is greater than can possibly result from gravity, since a body can never acquire, by gravity alone, a velocity greater than about seven miles per second. Some other cause, beside gravity, must therefore act, in order to give the meteors so great an apparent velocity.

In the fourth place, *the meteors fell towards the earth in straight lines, and in directions which, within considerable distances, were nearly parallel with each other.* The courses are inferred to have been in *straight lines*, because no others could have appeared to spectators in different situations to have described arcs of great circles. In order to be projected into the arc of a great circle, the line of descent must be in a plane passing through the eye of the spectator; and the intersection of such planes, passing through the eyes of different spectators, must be straight lines. The lines of direction are inferred to have been *parallel*, on account of their apparent radiation from one point, that being the vanishing point of parallel lines. This may appear to you a little paradoxical, to infer that lines are parallel, because they *diverge* from one and the same point; but it is a well-known principle of perspective, that parallel lines, when continued to a great distance from the eye, appear to converge towards the remoter end. You may observe this in two long rows of trees, or of street lamps.

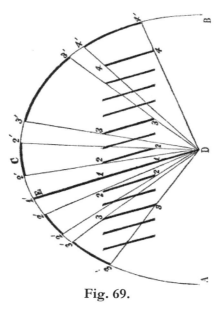

Fig. 69.

Some idea of the manner in which the meteors fell, and of the reason of their apparent radiation from a common point, may be gathered from the annexed diagram. Let A B C, Fig. 69, represent the vault of the sky, the centre of which, D, being the place of the spectator. Let 1, 2, 3, &c., represent parallel lines directed towards the earth. A luminous body descending through 1′ 1, coinciding with the line D E, coincident with the axis of vision, (or the line drawn from the meteoric body to the eye,) would appear stationary all the while at 1′, because distant bodies always appear stationary when they are moving either directly towards us or directly from us. A body descending through 2 2, would seem to describe the short arc 2′ 2′, appearing to move on the concave of the sky between the lines drawn from the eye to the two extremities of its line of motion; and, for a similar reason, a body descending through 3 3, would appear to describe the larger arc 3′ 3′. Hence, those meteors which fell nearer to the axis of vision, would describe shorter arcs, and move slower, while those which were further from the axis and nearer the horizon would appear to describe longer arcs, and to move with greater velocity; the meteors would all seem to radiate from a common centre, namely, the point where the axis of vision met the celestial vault; and if any meteor chanced to move directly in the line of vision, it would be seen as a luminous body, stationary, for a few seconds, at the centre of radiation. To see how exactly the facts, as observed, corresponded to these inferences, derived from the supposition that the meteors moved in *parallel lines*, take the following description, as given immediately after the occurrence, by Professor Twining. "In the vicinity of the radiant point, a few star-like bodies were observed, possessing very little motion, and leaving very little length of trace. Further off, the motions were more rapid and the traces longer; and most rapid of all, and longest in their traces, were those which originated but a few degrees above the horizon, and descended down to it."

In the fifth place, had the meteors come from a point twenty-two hundred and thirty-eight miles from the earth, and derived their apparent velocity from gravity alone, then it would be found, by a very easy calculation, that their actual velocity was about four miles per second; but, as already intimated, the velocity observed was estimated much greater than could be accounted for on these principles; not less, indeed, than fourteen miles per second, and, in some instances, much greater even than this. The motion of the earth in its orbit is about nineteen miles per second; and the most reasonable supposition we can make, at present, to account for the great velocity of the meteors, is, that they derived a relative motion from the earth's passing rapidly by them,—a supposition which is countenanced by the fact that they generally tended *westward* contrary to the earth's motion in its orbit.

In the sixth place, *the meteors consisted of combustible matter, and took fire, and were consumed, in traversing the atmosphere.* That these bodies underwent combustion, we had the direct evidence of the senses, inasmuch as we saw them burn. That they took fire in the *atmosphere*, was inferred from the fact that they were not luminous in their original situations in space, otherwise, we should have seen the body from which they emanated; and had they been luminous before reaching the atmosphere, we should have seen them for a much longer period than they were in sight, as they must have occupied a considerable time in descending towards the earth from so great a distance, even at the rapid rate at which they travelled. The immediate consequence of the prodigious velocity with which the meteors fell into the atmosphere must be a powerful condensation of the air before them, retarding their progress, and producing, by a sudden compression of the air, a great evolution of heat. There is a little instrument called the *air-match*, consisting of a piston and cylinder, like a syringe, in which we strike a light by suddenly forcing down the piston upon the air below. As the air cannot escape, it is suddenly compressed, and gives a spark sufficient to light a piece of tinder at the bottom of the cylinder. Indeed, it is a well-known fact, that, whenever air is suddenly and forcibly compressed, heat is elicited; and, if by such a compression as may be given by the hand in the air-match, heat is evolved sufficient to fire tinder, what must be the heat evolved by the motion of a large body in the atmosphere, with a velocity so immense. It is common to resort to electricity as the agent which produces the heat and light of shooting stars; but even were electricity competent to produce this effect, its presence, in the case before us, is not proved; and its agency is unnecessary, since so swift a motion of the meteors themselves, suddenly condensing the air before them, is both a known and adequate cause of an intense light and heat. A combustible body falling into the atmosphere, under such circumstances, would become speedily ignited, but could not burn freely, until it became enveloped in air of greater density; but, on reaching the lower portions of the atmosphere, it would burn with great rapidity.

In the seventh place, *some of the larger meteors must have been bodies of great size*. According to the testimony of various individuals, in different parts of the United States, a few fire-balls appeared as large as the full moon. Dr. Smith, (then of North Carolina, but since surgeon-general of the Texian army,) who was travelling all night on professional business, describes one which he saw in the following terms: "In size it appeared somewhat larger than the full moon rising. I was startled by the splendid light in which the surrounding scene was exhibited, rendering even small objects quite visible; but I heard no noise, although every sense seemed to be suddenly aroused, in sympathy with the violent impression on the sight." This description implies not only that the body was very large, but that it was at a considerable distance from the spectator. Its actual size will depend upon the distance; for, as it appeared under the same angle as the moon, its diameter will bear the same ratio to the moon's, as its distance bears to the moon's distance. We could, therefore, easily ascertain how large it was, provided we could find how far it was from the observer. If it was one hundred and ten miles distant, its diameter was one mile, and in the same proportion for a greater or less distance; and, if only at the distance of one mile, its diameter was forty-eight feet. For a moderate estimate, we will suppose it to have been twenty-two miles off; then its diameter was eleven hundred and fifty-six feet. Upon every view of the case, therefore, it must be admitted, that these were bodies of great size, compared with other objects which traverse the atmosphere. We may further infer the great magnitude of some of the meteors, from the dimensions of the trains, or clouds, which resulted from their destruction. These often extended over several degrees, and at length were borne along in the direction of the wind, exactly in the manner of a small cloud.

It was an interesting problem to ascertain, if possible, the height above the earth at which these fire-balls exploded, or resolved themselves into a cloud of smoke. This would be an easy task, provided we could be certain that two or more distant observers could be sure that both saw the same meteor; for as each would refer the place of explosion, or the position of the cloud that resulted from it, to a different point of the sky, a parallax would thus be obtained, from which the height might be determined. The large meteor which is mentioned in my account of the shower, (see page 348,) as having exploded near the star Capella, was so peculiar in its appearance, and in the form and motions of the small cloud which resulted from its combustion, that it was noticed and distinguished by a number of observers in distant parts of the country. All described the meteor as exhibiting, substantially, the same peculiarities of appearance; all agreed very nearly in the time of its occurrence; and, on drawing lines, to represent the course and direction of the place where it exploded to the view of each of the observers respectively, these lines met in nearly one and the same point, and that was over the place where it was seen in the zenith. Little doubt, therefore, could remain, that all saw the same body; and on ascertaining, from a comparison of their observations, the amount of parallax, and thence deducing its height,—a task which was ably executed by Professor Twining,—the following results were obtained: that this meteor, and probably all the meteors, entered the atmosphere with a velocity not less, but perhaps greater, than *fourteen miles in a second*; that they became luminous many miles from the earth,—in this case, over *eighty miles*; and became extinct high above the surface,—in this case, nearly *thirty miles*.

In the eighth place, *the meteors were combustible bodies, and were constituted of light and transparent materials.* The fact that they burned is sufficient proof that they belonged to the class of *combustible* bodies; and they must have been composed of very *light materials*, otherwise their momentum would have been sufficient to enable them to make their way through the atmosphere to the surface of the earth. To compare great things with small, we may liken them to a wad discharged from a piece of artillery, its velocity being supposed to be increased (as it may be) to such a degree, that it shall take fire as it moves through the air. Although it would force its way to a great distance from the gun, yet, if not consumed too soon, it would at length be stopped by the resistance of the air. Although it is supposed that the meteors did in fact slightly disturb the atmospheric equilibrium, yet, had they been constituted of dense matter, like meteoric stones, they would doubtless have disturbed it vastly more. Their own momentum would be lost only as it was imparted to the air; and had such a number of bodies,— some of them quite large, perhaps a mile in diameter, and entering the atmosphere with a velocity more than forty times the greatest velocity of a cannon ball,—had they been composed of dense, ponderous matter, we should have had appalling evidence of this fact, not only in the violent winds which they would have produced in the atmosphere, but in the calamities they would have occasioned on the surface of the earth. The meteors were *transparent* bodies; otherwise, we cannot conceive why the body from which they emanated was not distinctly visible, at least by reflecting the light of the sun. If only the meteors which were known to fall towards the earth had been collected and restored to their original connexion in space, they would have composed a body of great extent; and we cannot imagine a body of such dimensions, under such circumstances, which would not be visible, unless formed of highly transparent materials. By these unavoidable inferences respecting the kind of matter of which the meteors were composed, we are unexpectedly led to recognise a body bearing, in its constitution, a strong analogy to comets, which are also composed of exceedingly light and transparent, and, as there is much reason to believe, of combustible matter.

We now arrive at the final inquiry, *what relations did the body which afforded the meteoric shower sustain to the earth?* Was it of the nature of a satellite, or terrestrial comet, that revolves around the earth as its centre of motion? Was it a collection of nebulous, or cometary matter, which the earth encountered in its annual progress? or was it a comet, which chanced at this time to be pursuing its path along with the earth, around their common centre of motion? It could not have been of the nature of a satellite to the earth, (or one of those bodies which are held by some to afford the meteoric stones, which sometimes fall to the earth from huge meteors that traverse the atmosphere,) because it remained so long stationary with respect to the earth. A body so near the earth as meteors of this class are known to be, could not remain apparently stationary among the stars for a moment; whereas the body in question occupied the same position, with hardly any perceptible variation, for at least two hours. Nor can we suppose that the earth, in its annual progress, came into the vicinity of a *nebula*, which was either stationary, or wandering lawless through space. Such a collection of matter could not remain stationary within the solar system, in an insulated state, for, if not prevented by a motion of its own, or by the attraction of some nearer body, it would have proceeded directly towards the sun; and had it been in motion in any other direction than that in which the earth was moving, it would soon have been separated from the earth; since, during the eight hours, while the meteoric shower was visible, the earth moved in its orbit through the space of nearly five hundred and fifty thousand miles.

The foregoing considerations conduct us to the following train of reasoning. First, if all the meteors which fell on the morning of November

13, 1833, had been collected and restored to their original connexion in space, they would of themselves have constituted a nebulous body of great extent; but we have reason to suppose that they, in fact, composed but a small part of the mass from which they emanated, since, after the loss of so much matter as proceeded from it in the great meteoric shower of 1799, and in the several repetitions of it that preceded the year 1833, it was still capable of affording so copious a shower on that year; and similar showers, more limited in extent, were repeated for at least five years afterwards. We are therefore to regard the part that descended only as *the extreme portions of a body or collection of meteors, of unknown extent, existing in the planetary spaces.*

Secondly, since the earth fell in with this body in the same part of its orbit, for several years in succession, it must either have remained there while the earth was performing its whole revolution around the sun, or it must itself have had a revolution, as well as the earth. But I have already shown that it could not have remained stationary in that part of space; therefore, *it must have had a revolution around the sun.*

Thirdly, its period of revolution must have either been greater than the earth's, equal to it, or less. It could not have been greater, for then the two bodies could not have been together again at the end of the year, since the meteoric body would not have completed its revolution in a year. Its period might obviously be the same as the earth's, for then they might easily come together again after one revolution of each; although their orbits might differ so much in shape as to prevent their being together at any intermediate point. But the period of the body might also be less than that of the earth, provided it were some *aliquot part of a year*, so as to revolve just twice, or three times, for example, while the earth revolves once. Let us suppose that the period is one third of a year. Then, since we have given the periodic times of the two bodies, and the major axis of the orbit of one of them, namely, of the earth, we can, by Kepler's law, find the major axis of the other orbit; for the square of the earth's periodic time 1^2 is to the square of the body's time $(\frac{1}{3})^2$ as the cube of the major axis of the earth's orbit is to the cube of the major axis of the orbit in question. Now, the three first terms of this proportion are known, and consequently, it is only to solve a case in the simple rule of three, to find the term required. On making the calculation, it is found, that the supposition of a periodic time of only one third of a year gives an orbit of insufficient length; the whole major axis would not reach from the sun to the earth; and consequently, a body revolving in it could never come near to the earth. On making trial of six months, we obtain an orbit which satisfies the conditions, being such as is represented by the diagram on page 362, Fig. 69´, where the outer circle denotes the earth's orbit, the sun being in the centre, and the inner ellipse denotes the path of the meteoric body. The two bodies are together at the top of the figure, being the place of the meteoric body's aphelion on the thirteenth of November, and the figures 10, 20, &c., denote the relative positions of the earth and the body for every ten days, for a period of six months, in which time the body would have returned to its aphelion.

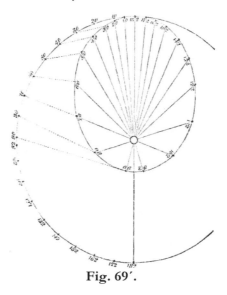

Fig. 69'.

Such would be the relation of the body that affords the meteoric shower of November, provided its revolution is accomplished in six months; but it is still somewhat uncertain whether the period be half a year or a year; it must be one or the other.

If we inquire, now, why the meteors always appear to radiate from a point in the constellation Leo, recollecting that this is the point to which the body is projected among the stars, the answer is, that this is the very point towards which the earth is moving in her orbit at that time; so that if, as we have proved, the earth passed through or near a nebulous body on the thirteenth of November, that body must necessarily have been projected into the constellation Leo, else it could not have lain directly in her path. I consider it therefore as established by satisfactory proof, that the meteors of November thirteenth emanate from a nebulous or cometary body, revolving around the sun, and coming so near the earth at that time that the earth passes through its *skirts*, or extreme portions, and thus attracts to itself some portions of its matter, giving to the meteors a greater velocity than could be imparted by gravity alone, in consequence of passing rapidly by them.

All these conclusions were made out by a process of reasoning strictly inductive, without supposing that the meteoric body itself had ever been seen. But there are some reasons for believing that we do actually see it, and that it is no other than that mysterious appearance long known under the name of the *zodiacal light*. This is a faint light, which at certain seasons of the year appears in the west after evening twilight, and at certain other seasons appears in the east before the dawn, following or preceding the track of the sun in a triangular figure, with its broad base next to the sun, and its vertex reaching to a greater or less distance, sometimes more than ninety degrees from that luminary. You may obtain a good view of it in February or March, in the west, or in October, in the morning sky. The various changes which this light undergoes at different seasons of the year are such as to render it probable, to my mind, that this is the very body which affords the meteoric showers; its extremity coming, in November, within the sphere of the earth's attraction. But, as the arguments for the existence of a body in the planetary regions, which affords these showers, were drawn without the least reference to the zodiacal light, and are good, should it finally be proved that this light has no connexion with them, I will not occupy your attention with the discussion of this point, to the exclusion of topics which will probably interest you more.

It is perhaps most probable, that the meteoric showers of August and December emanate from the same body. I know of nothing repugnant to this conclusion, although it has not yet been distinctly made out. Had the periods of the earth and of the meteoric body been so adjusted to each other that the latter was contained an exact even number of times in the former; that is, had it been *exactly* either a year or half a year; then we might expect a similar recurrence of the meteoric shower every year; but only a slight variation in such a proportion between the two periods would occasion the repetition of the shower for a few years in succession, and then an intermission of them, for an unknown length of time, until the two bodies were brought into the same relative situation as before. Disturbances, also, occasioned by the action of Venus and Mercury, might wholly subvert this numerical relation, and increase or diminish the probability of a repetition of the phenomenon. Accordingly, from the year 1830, when the meteoric shower of November was first observed, until 1833, there was a regular increase of the exhibition; in 1833, it came to its maximum; and after that time it was repeated upon a constantly diminishing scale, until 1838, since which time it has not been observed. Perhaps ages may roll away before the world will be again surprised and delighted with a display of celestial fire-works equal to that of the morning of November 13, 1833.

LETTER XXVIII.

FIXED STARS.

——"O, majestic Night! Nature's great ancestor!
Day's elder born, And fated to survive the transient sun!
By mortals and immortals seen with awe!
A starry crown thy raven brow adorns,
An azure zone thy waist; clouds, in heaven's
loom Wrought, through varieties of shape and shade,
In ample folds of drapery divine,
Thy flowing mantle form; and heaven throughout
Voluminously pour thy pompous train."—*Young.*

SINCE the solar system is but one among a myriad of worlds which astronomy unfolds, it may appear to you that I have dwelt too long on so diminutive a part of creation, and reserved too little space for the other systems of the universe. But however humble a province our sun and planets compose, in the vast empire of Jehovah, yet it is that which most concerns us; and it is by the study of the laws by which this part of creation is governed, that we learn the secrets of the skies.

Until recently, the observation and study of the phenomena of the solar system almost exclusively occupied the labors of astronomers. But Sir William Herschel gave his chief attention to the *sidereal heavens*, and opened new and wonderful fields of discovery, as well as of speculation. The same subject, has been prosecuted with similar zeal and success by his son, Sir John Herschel, and Sir James South, in England, and by Professor Struve, of Dorpat, until more has been actually achieved than preceding astronomers had ventured to conjecture. A limited sketch of these wonderful discoveries is all that I propose to offer you.

The fixed stars are so called, because, to common observation, they always maintain the same situations with respect to one another. The stars are classed by their apparent *magnitudes*. The whole number of magnitudes recorded are *sixteen*, of which the first six only are visible to the naked eye; the rest are *telescopic stars*. These magnitudes are not determined by any very definite scale, but are merely ranked according to their relative degrees of brightness, and this is left in a great measure to the decision of the eye alone. The brightest stars, to the number of fifteen or twenty, are considered as stars of the first magnitude; the fifty or sixty next brightest, of the second magnitude; the next two hundred, of the third magnitude;

and thus the number of each class increases rapidly, as we descend the scale, so that no less than fifteen or twenty thousand are included within the first seven magnitudes.

The stars have been grouped in *constellations* from the most remote antiquity; a few, as Orion, Bootes, and Ursa Major, are mentioned in the most ancient writings, under the same names as they bear at present. The names of the constellations are sometimes founded on a supposed resemblance to the objects to which they belong; as the Swan and the Scorpion were evidently so denominated from their likeness to those animals; but in most cases, it is impossible for us to find any reason for designating a constellation by the figure of the animal or hero which is employed to represent it. These representations were probably once blended with the fables of pagan mythology. The same figures, absurd as they appear, are still retained for the convenience of reference; since it is easy to find any particular star, by specifying the part of the figure to which it belongs; as when we say, a star is in the neck of Taurus, in the knee of Hercules, or in the tail of the Great Bear. This method furnishes a general clue to its position; but the stars belonging to any constellation are distinguished according to their apparent magnitudes, as follows: First, by the Greek letters, Alpha, Beta, Gamma, &c. Thus, *Alpha Orionis* denotes the largest star in Orion; *Beta Andromedæ* the second star in Andromeda; and *Gamma Leonis*, the third brightest star in the Lion. When the number of the Greek letters is insufficient to include all the stars in a constellation, recourse is had to the letters of the Roman alphabet, a, b, c, &c.; and in all cases where these are exhausted the final resort is to numbers. This is evidently necessary, since the largest constellations contain many hundreds or even thousands of stars. *Catalogues* of particular stars have also been published, by different astronomers, each author numbering the individual stars embraced in his list according to the places they respectively occupy in the catalogue. These references to particular catalogues are sometimes entered on large celestial globes. Thus we meet with a star marked 84 H., meaning that this is its number in Herschel's catalogue; or 140 M., denoting the place the star occupies in the catalogue of Mayer.

The earliest catalogue of the stars was made by Hipparchus, of the Alexandrian school, about one hundred and forty years before the Christian era. A new star appearing in the firmament, he was induced to count the stars, and to record their positions, in order that posterity might be able to judge of the permanency of the constellations. His catalogue contains all that were conspicuous to the naked eye in the latitude of Alexandria, being one thousand and twenty-two. Most persons, unacquainted with the actual number of the stars which compose the visible firmament, would suppose it to be much greater than this; but it is found that the catalogue of

Hipparchus embraces nearly all that can now be seen in the same latitude; and that on the equator, where the spectator has both the northern and southern hemispheres in view, the number of stars that can be counted does not exceed three thousand. A careless view of the firmament in a clear night gives us the impression of an infinite number of stars; but when we begin to count them, they appear much more sparsely distributed than we supposed, and large portions of the sky appear almost destitute of stars.

By the aid of the telescope, new fields of stars present themselves, of boundless extent; the number continually augmenting, as the powers of the telescope are increased. Lalande, in his 'Histoire Celeste,' has registered the positions of no less than fifty thousand; and the whole number visible in the largest telescopes amounts to many millions.

When you look at the firmament on a clear Autumnal or Winter evening, it appears so thickly studded with stars, that you would perhaps imagine that the task of learning even the brightest of them would be almost hopeless. Let me assure you, this is all a mistake. On the contrary, it is a very easy task to become acquainted with the names and positions of the stars of the first magnitude, and of the leading constellations. If you will give a few evenings to the study, you will be surprised to find, both how rapidly you can form these new acquaintances, and how deeply you will become interested in them. I would advise you, at first, to obtain, for an evening or two, the assistance of some friend who is familiar with the stars, just to point out a few of the most conspicuous constellations. This will put you on the track, and you will afterwards experience no difficulty in finding all the constellations and stars that are particularly worth knowing; especially if you have before you a map of the stars, or, what is much better, a celestial globe. It is a pleasant evening recreation for a small company of young astronomers to go out together, and learn one or two constellations every favorable evening, until the whole are mastered. If you have a celestial globe, *rectify* it for the evening; that is, place it in such a position, that the constellations shall be seen on it in the same position with respect to the horizon, that they have at that moment in the sky itself. To do this, I first elevate the north pole until the number of degrees on the brass meridian from the pole to the horizon corresponds to my latitude, (forty-one degrees and eighteen minutes.) I then find the sun's place in the ecliptic, by looking for the day of the month on the broad horizon, and against it noting the corresponding sign and degree. I now find the same sign and degree on the ecliptic itself, and bring that point to the brass meridian. As that will be the position of the sun at noon, I set the hour-index at twelve, and then turn the globe westward, until the index points to the given hour of the evening. If I now inspect the figures of the constellations, and then look upward at the firmament, I shall see that the latter are spread over the sky in the same manner as the pictures of them are painted on the globe. I will point out a few marks by which the leading constellations may be recognised; this will aid you in finding them, and you can afterwards learn the individual stars of a constellation, to any extent you please, by means of the globes or maps. Let us begin with the *Constellations of the Zodiac*, which, succeeding each other, as they do, in a known order, are most easily found.

Aries (*the Ram*) is a small constellation, known by two bright stars which form his head, *Alpha* and *Beta Arietis*. These two stars are about four degrees apart; and directly south of Beta, at the distance of one degree, is a smaller star, *Gamma Arietis*. It has been already intimated that the Vernal equinox probably was near the head of Aries, when the signs of the zodiac received their present names.

Taurus (*the Bull*) will be readily found by the seven stars, or *Pleiades*, which lie in his neck. The largest star in Taurus is *Aldebaran*, in the Bull's eye, a star of the first magnitude, of a reddish color, somewhat resembling the planet Mars. Aldebaran and four other stars, close together in the face of Taurus, compose the *Hyades*.

Gemini (*the Twins*) is known by two very bright stars, *Castor and Pollux*, five degrees asunder. Castor (the northern) is of the first, and Pollux of the second, magnitude.

Cancer (*the Crab*.) There are no large stars in this constellation, and it is regarded as less remarkable than any other in the zodiac. It contains, however, an interesting group of small stars, called *Præsepe*, or the nebula of Cancer, which resembles a comet, and is often mistaken for one, by persons unacquainted with the stars. With a telescope of very moderate powers this nebula is converted into a beautiful assemblage of exceedingly bright stars.

Leo (*the Lion*) is a very large constellation, and has many interesting members. *Regulus* (*Alpha Leonis*) is a star of the first magnitude, which lies directly in the ecliptic, and is much used in astronomical observations. North of Regulus, lies a semicircle of bright stars, forming a *sickle*, of which Regulus is the handle. *Denebola*, a star of the second magnitude, is in the Lion's tail, twenty-five degrees northeast of Regulus.

Virgo (*the Virgin*) extends a considerable way from west to east, but contains only a few bright stars. *Spica*, however, is a star of the first magnitude, and lies a little east of the place of the Autumnal equinox. Eighteen degrees eastward of Denebola, and twenty degrees north of Spica, is *Vindemiatrix*, in the arm of Virgo, a star of the third magnitude.

Libra (the Balance) is distinguished by three large stars, of which the two brightest constitute the beam of the balance, and the smallest forms the top or handle.

Scorpio (the Scorpion) is one of the finest of the constellations. His head is formed of five bright stars, arranged in the arc of a circle, which is crossed in the centre by the ecliptic nearly at right angles, near the brightest of the five, *Beta Scorpionis*. Nine degrees southeast of this is a remarkable star of the first magnitude, of a reddish color, called *Cor Scorpionis*, or *Antares*. South of this, a succession of bright stars sweep round towards the east, terminating in several small stars, forming the tail of the Scorpion.

Sagittarius (the Archer.) Northeast of the tail of the Scorpion are three stars in the arc of a circle, which constitute the *bow* of the Archer, the central star being the brightest, directly west of which is a bright star which forms the *arrow*.

Capricornus (the Goat) lies northeast of Sagittarius, and is known by two bright stars, three degrees apart, which form the head.

Aquarius (the Water-Bearer) is recognised by two stars in a line with *Alpha Capricorni*, forming the shoulders of the figure. These two stars are ten degrees apart; and three degrees southeast is a third star, which, together with the other two, make an acute triangle, of which the westernmost is the vertex.

Pisces (the Fishes) lie between Aquarius and Aries. They are not distinguished by any large stars, but are connected by a series of small stars, that form a crooked line between them. *Piscis Australis*, the Southern Fish, lies directly below Aquarius, and is known by a single bright star far in the south, having a declination of thirty degrees. The name of this star is *Fomalhaut*, and it is much used in astronomical measurements.

The constellations of the zodiac, being first well learned, so as to be readily recognised, will facilitate the learning of others that lie north and south of them. Let us, therefore, next review the principal *Northern Constellations*, beginning north of Aries, and proceeding from west to east.

Andromeda is characterized by three stars of the second magnitude, situated in a straight line, extending from west to east. The middle star is about seventeen degrees north of Beta Arietis. It is in the girdle of Andromeda, and is named *Mirach*. The other two lie at about equal distances, fourteen degrees west and east of Mirach. The western star, in the head of Andromeda, lies in the equinoctial colure. The eastern star, *Alamak*, is situated in the foot.

Perseus lies directly north of the Pleiades, and contains several bright stars. About eighteen degrees from the Pleiades is *Algol*, a star of the second magnitude, in the head of Medusa, which forms a part of the figure; and nine degrees northeast of Algol is *Algenib*, of the same magnitude, in the back of Perseus. Between Algenib and the Pleiades are three bright stars, at nearly equal intervals, which compose the right leg of Perseus.

Auriga (*the Wagoner*) lies directly east of Perseus, and extends nearly parallel to that constellation, from north to south. *Capella*, a very white and beautiful star of the first magnitude, distinguishes this constellation. The feet of Auriga are near the Bull's horns.

The *Lynx* comes next, but presents nothing particularly interesting, containing no stars above the fourth magnitude.

Leo Minor consists of a collection of small stars north of the sickle in Leo, and south of the Great Bear. Its largest star is only of the third magnitude.

Coma Berenices is a cluster of small stars, north of Denebola, in the tail of the Lion, and of the head of Virgo. About twelve degrees directly north of Berenice's hair, is a single bright star, called *Cor Caroli*, or Charles's Heart.

Bootes, which comes next, is easily found by means of *Arcturus*, a star of the first magnitude, of a reddish color, which is situated near the knee of the figure. Arcturus is accompanied by three small stars, forming a triangle a little to the southwest. Two bright stars, *Gamma* and *Delta Bootis*, form the shoulders, and *Beta*, of the third magnitude, is in the head, of the figure.

Corona Borealis, (*the Crown*,) which is situated east of Bootes, is very easily recognised, composed as it is of a semicircle of bright stars. In the centre of the bright crown is a star of the second magnitude, called *Gemma*: the remaining stars are all much smaller.

Hercules, lying between the Crown on the west and the Lyre on the east, is very thickly set with stars, most of which are quite small. This constellation covers a great extent of the sky, especially from north to south, the head terminating within fifteen degrees of the equator, and marked by a star of the third magnitude, called *Ras Algethi*, which is the largest in the constellation.

Ophiucus is situated directly south of Hercules, extending some distance on both sides of the equator, the feet resting on the Scorpion. The head terminates near the head of Hercules, and, like that, is marked by a bright star within five degrees of *Alpha Herculis* Ophiucus is represented as holding in his hands the *Serpent*, the head of which, consisting of three bright stars, is situated a little south of the Crown. The folds of the serpent will be easily followed by a succession of bright stars, which extend a great way to the east.

Aquila (*the Eagle*) is conspicuous for three bright stars in its neck, of which the central one, *Altair*, is a very brilliant white star of the first magnitude. *Antinous* lies directly south of the Eagle, and north of the head of Capricornus.

Delphinus (*the Dolphin*) is a small but beautiful constellation, a few degrees east of the Eagle, and is characterized by four bright stars near to one another, forming a small rhombic square. Another star of the same magnitude, five degrees south, makes the tail.

Pegasus lies between Aquarius on the southwest and Andromeda on the northeast. It contains but few large stars. A very regular square of bright stars is composed of *Alpha Andromedæ* and the three largest stars in Pegasus; namely, *Scheat*, *Markab*, and *Algenib*. The sides composing this square are each about fifteen degrees. Algenib is situated in the equinoctial colure.

We may now review the *Constellations which surround the north pole*, within the circle of perpetual apparition.

Ursa Minor (*the Little Bear*) lies nearest the pole. The pole-star, *Polaris*, is in the extremity of the tail, and is of the third magnitude. Three stars in a straight line, four degrees or five degrees apart, commencing with the pole-star, lead to a trapezium of four stars, and the whole seven form together a *dipper*,—the trapezium being the body and the three stars the handle.

Ursa Major (*the Great Bear*) is situated between the pole and the Lesser Lion, and is usually recognised by the figure of a larger and more perfect dipper which constitutes the hinder part of the animal. This has also seven stars, four in the body of the Dipper and three in the handle. All these are stars of much celebrity. The two in the western side of the Dipper, Alpha and Beta, are called *Pointers*, on account of their always being in a right line with the pole-star, and therefore affording an easy mode of finding that. The first star in the tail, next the body, is named *Alioth*, and the second, *Mizar*. The head of the Great Bear lies far to the westward of the Pointers, and is composed of numerous small stars; and the feet are severally composed of two small stars very near to each other.

Draco (*the Dragon*) winds round between the Great and the Little Bear; and, commencing with the tail, between the Pointers and the pole-star, it is easily traced by a succession of bright stars extending from west to east.

Passing under Ursa Minor, it returns westward, and terminates in a triangle which forms the head of Draco, near the feet of Hercules, northwest of Lyra. *Cepheus* lies eastward of the breast of the Dragon, but has no stars above the third magnitude.

Cassiopeia is known by the figure of a *chair*, composed of four stars which form the legs, and two which form the back. This constellation lies between Perseus and Cepheus, in the Milky Way.

Cygnus (*the Swan*) is situated also in the Milky Way, some distance southwest of Cassiopeia, towards the Eagle. Three bright stars, which lie along the Milky Way, form the body and neck of the Swan, and two others, in a line with the middle one of the three, one above and one below, constitute the wings. This constellation is among the few that exhibit some resemblance to the animals whose names they bear.

Lyra (*the Lyre*) is directly west of the Swan, and is easily distinguished by a beautiful white star of the first magnitude, *Alpha Lyræ*.

The *Southern Constellations* are comparatively few in number. I shall notice only the Whale, Orion, the Greater and Lesser Dog, Hydra, and the Crow.

Cetus (*the Whale*) is distinguished rather for its extent than its brilliancy, reaching as it does through forty degrees of longitude, while none of its stars, except one, are above the third magnitude. *Menkar* (*Alpha Ceti*) in the mouth, is a star of the second magnitude; and several other bright stars, directly south of Aries, make the head and neck of the Whale. *Mira*, (*Omicron Ceti*,) in the neck of the Whale, is a variable star.

Orion is one of the largest and most beautiful of the constellations, lying southeast of Taurus. A cluster of small stars forms the head; two large stars, *Betalgeus* of the first and *Bellatrix* of the second magnitude, make the shoulders; three more bright stars compose the buckler, and three the sword; and *Rigel*, another star of the first magnitude, makes one of the feet. In this constellation there are seventy stars plainly visible to the naked eye, including two of the first magnitude, four of the second, and three of the third.

Canis Major lies southeast of Orion, and is distinguished chiefly by its containing the largest of the fixed stars, *Sirius*.

Canis Minor, a little north of the equator, between Canis Major and Gemini, is a small constellation, consisting chiefly of two stars, of which, *Procyon* is of the first magnitude.

Hydra has its head near Procyon, consisting of a number of stars of ordinary brightness. About fifteen degrees southeast of the head is a star of the second magnitude, forming the heart, (*Cor Hydræ,*) and eastward of this is a long succession of stars of the fourth and fifth magnitudes, composing the body and tail, and reaching a few degrees south of Spica Virginis.

Corvus (the Crow) is represented as standing on the tail of Hydra. It consists of small stars, only three of which are as large as the third magnitude.

In assigning the places of individual stars, I have not aimed at great precision; but such a knowledge as you will acquire of the constellations and larger stars, by nothing more even than you can obtain from the foregoing sketch, will not only add greatly to the interest with which you will ever afterwards look at the starry heavens, but it will enable you to locate any phenomenon that may present itself in the nocturnal sky, and to understand the position of any object that may be described, by assigning its true place among the stars; although I hope you will go much further than this mere outline, in cultivating an actual acquaintance with the stars. Leaving, now, these great divisions of the bodies of the firmament, let us ascend to the next order of stars, composing CLUSTERS.

In various parts of the nocturnal heavens are seen large groups which, either by the naked eye, or by the aid of the smallest telescope, are perceived to consist of a great number of small stars. Such are the Pleiades, Coma Berenices, and Præsepe, or the Bee-hive, in Cancer. The *Pleiades*, or Seven Stars, as they are called, in the neck of Taurus, is the most conspicuous cluster. When we look *directly* at this group, we cannot distinguish more than six stars; but by turning the eye *sideways* upon it, we discover that there are many more; for it is a remarkable fact that indirect vision is far more delicate than direct. Thus we can see the zodiacal light or a comet's tail much more distinctly and better defined, if we fix one eye on a part of the heavens at some distance and turn the other eye obliquely upon the object, than we can by looking directly towards it. Telescopes show the Pleiades to contain fifty or sixty stars, crowded together, and apparently insulated from the other parts of the heavens. *Coma Berenices* has fewer stars, but they are of a larger class than those which compose the Pleiades. The *Bee-hive*, or Nebula of Cancer, as it is called, is one of the finest objects of this kind for a small telescope, being by its aid converted into a rich congeries of shining points. The head of Orion affords an example of another cluster, though less remarkable than those already mentioned. These clusters are pleasing objects to the telescope; and since a common spyglass will serve to give a distinct view of most of them, every one may have the power of taking the view. But we pass, now, to the third order of stars, which present themselves much more obscurely to the gaze of the astronomer, and require large instruments for the full developement of their wonderful organization. These are the NEBULÆ.

Figures 70, 71, 72, 73. CLUSTERS OF STARS AND NEBULÆ.

Nebulæ are faint misty appearances which are dimly seen among the stars, resembling comets, or a speck of fog. They are usually resolved by the telescope into myriads of small stars; though in some instances, no powers of the telescope have been found sufficient thus to resolve them. The *Galaxy* or Milky Way, presents a continued succession of large nebulas. The telescope reveals to us innumerable objects of this kind. Sir William Herschel has given catalogues of two thousand nebulæ, and has shown that the nebulous matter is distributed through the immensity of space in quantities inconceivably great, and in separate parcels, of all shapes and sizes, and of all degrees of brightness between a mere milky appearance and the condensed light of a fixed star. In fact, more distinct nebulæ have been hunted out by the aid of telescopes than the whole number of stars visible to the naked eye in a clear Winter's night. Their appearances are extremely diversified. In many of them we can easily distinguish the individual stars; in those apparently more remote, the interval between the stars diminishes, until it becomes quite imperceptible; and in their faintest aspect they dwindle to points so minute, as to be appropriately denominated *star-dust*. Beyond this, no stars are distinctly visible, but only streaks or patches of milky light. The diagram facing page 379 represents a magnificent nebula in the Galaxy. In objects so distant as the fixed stars, any apparent interval must denote an immense space; and just imagine yourself situated any where within the grand assemblage of stars, and a firmament would expand itself over your head like that of our evening sky, only a thousand times more rich and splendid.

Many of the nebulæ exhibit a tendency towards a globular form, and indicate a rapid condensation towards the centre. This characteristic is exhibited in the forms represented in Figs. 70 and 71. We have here two specimens of nebulæ of the nearer class, where the stars are easily discriminated. In Figs. 72 and 73 we have examples of two others of the remoter kind, one of which is of the variety called *star-dust*. These wonderful objects, however, are not confined to the spherical form, but exhibit great varieties of figure. Sometimes they appear as ovals; sometimes they are shaped like a fan; and the unresolvable kind often affect the most fantastic forms. The opposite diagram, Fig. 74, as well as the preceding, affords a specimen of these varieties, as given in Professor Nichols's 'Architecture of the Heavens,' where they are faithfully copied from the papers of Herschel, in the 'Philosophical Transactions.'

Sir John Herschel has recently returned from a residence of five years at the Cape of Good Hope, with the express view of exploring the hidden treasures of the southern hemisphere. The kinds of nebulæ are in general similar to those of the northern hemisphere, and the forms are equally various and singular. The *Magellan Clouds*, two remarkable objects seen among the stars of that hemisphere, and celebrated among navigators, appeared to the great telescope of Herschel (as we are informed by Professor Nichols) no longer as simple milky spots, or permanent light flocculi of cloud, as they appear to the unassisted eye, but shone with inconceivable splendor. The *Nubecula Major*, as the larger object is called, is a congeries of clusters of stars, of irregular form, globular clusters and nebulæ of various magnitudes and degrees of condensation, among which is interspersed a large portion of irresolvable nebulous matter, which may be, and probably is, star-dust, but which the power of the twenty-feet telescope shows only as a general illumination of the field of view, forming a bright ground on which the other objects are scattered. The *Nubecula Minor* (the lesser cloud) exhibited appearances similar, though inferior in degree.

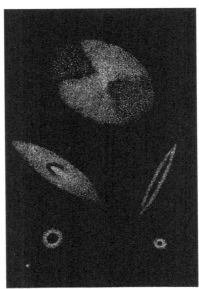

Figure 74. VARIOUS FORMS OF NEBULÆ.

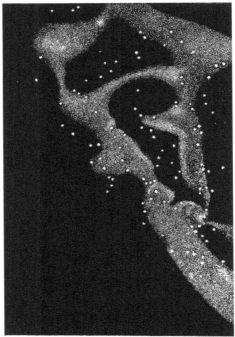

Figure 75. A NEBULA IN THE MILKY WAY.

It is a grand idea, first conceived by Sir William Herschel, and generally adopted by astronomers, that the whole Galaxy, or Milky Way, is nothing else than a nebula, and appears so extended, merely because it happens to be that particular nebula to which we belong. According to this view, our sun, with his attendant planets and comets, constitutes but a single star of the Galaxy, and our firmament of stars, or visible heavens, is composed of the stars of *our* nebula alone. An inhabitant of any of the other nebulæ would see spreading over him a firmament equally spacious, and in some cases inconceivably more brilliant.

It is an exalted spectacle to travel over the Galaxy in a clear night, with a powerful telescope, with the heart full of the idea that every star is a world. Sir William Herschel, by counting the stars in a single field of his telescope, estimated that fifty thousand had passed under his review in a zone two degrees in breadth, during a single hour's observation. Notwithstanding the apparent contiguity of the stars which crowd the Galaxy, it is certain that their mutual distances must be inconceivably great.

It is with some reluctance that I leave, for the present, this fairy land of astronomy; but I must not omit, before bringing these Letters to a conclusion, to tell you something respecting other curious and interesting objects to be found among the stars.

VARIABLE STARS are those which undergo a periodical change of brightness. One of the most remarkable is the star *Mira*, in the Whale, (*Omicron Ceti.*) It appears once in eleven months, remains at its greatest brightness about a fortnight, being then, on some occasions, equal to a star of the second magnitude. It then decreases about three months, until it becomes completely invisible, and remains so about five months, when it again becomes visible, and continues increasing during the remaining three months of its period.

Another very remarkable variable star is *Algol*, (*Beta Persei.*) It is usually visible as a star of the second magnitude, and continues such for two days and fourteen hours, when it suddenly begins to diminish in splendor, and in about three and a half hours is reduced to the fourth magnitude. It then begins again to increase, and in three and a half hours more is restored to its usual brightness, going through all its changes in less than three days. This remarkable law of variation appears strongly to suggest the revolution round it of some opaque body, which, when interposed between us and Algol, cuts off a large portion of its light. "It is," says Sir J. Herschel, "an indication of a high degree of activity in regions where, but for such evidences, we might conclude all lifeless. Our sun requires almost nine times this period to perform a revolution on its axis. On the other hand, the periodic time of an opaque revolving body, sufficiently large, which would produce a similar temporary obscuration of the sun, seen from a fixed star, would be less than fourteen hours." The duration of these periods is extremely various. While that of Beta Persei, above mentioned, is less than three days, others are more than a year; and others, many years.

TEMPORARY STARS are new stars, which have appeared suddenly in the firmament, and, after a certain interval, as suddenly disappeared, and returned no more. It was the appearance of a new star of this kind, one hundred and twenty-five years before the Christian era, that prompted Hipparchus to draw up a catalogue of the stars, the first on record. Such, also, was the star which suddenly shone out, A.D. 389, in the Eagle, as bright as Venus, and, after remaining three weeks, disappeared entirely. At other periods, at distant intervals, similar phenomena have presented themselves. Thus the appearance of a star in 1572 was so sudden, that Tycho Brahe, returning home one day, was surprised to find a collection of country people gazing at a star which he was sure did not exist half an hour before. It was then as bright as Sirius, and continued to increase until it surpassed Jupiter when brightest, and was visible at mid-day. In a month it began to diminish; and, in three months afterwards, it had entirely disappeared. It has been supposed by some that, in a few instances, the same star has returned, constituting one of the periodical or variable stars of a long period. Moreover, on a careful reexamination of the heavens, and a comparison of catalogues, many stars are now discovered to be missing.

DOUBLE STARS are those which appear single to the naked eye, but are resolved into two by the telescope; or, if not visible to the naked eye, are seen in the telescope so close together as to be recognised as objects of this class. Sometimes, three or more stars are found in this near connexion, constituting triple, or multiple stars. Castor, for example, when seen by the naked eye, appears as a single star, but in a telescope even of moderate powers, it is resolved into two stars, of between the third and fourth magnitudes, within five seconds of each other. These two stars are nearly of equal size; but more commonly, one is exceedingly small in comparison with the other, resembling a satellite near its primary, although in distance, in light, and in other characteristics, each has all the attributes of a star, and the combination, therefore, cannot be that of a planet with a satellite. In most instances, also, the distance between these objects is much less than five seconds; and, in many cases, it is less than one second. The extreme closeness, together with the exceeding minuteness, of most of the double stars, requires the best telescopes united with the most acute powers of observation. Indeed, certain of these objects are regarded as the severest *tests* both of the excellence of the instruments and of the skill of the observer. The diagram on page 382, Fig. 76, represents four double stars, as seen with appropriate magnifiers. No. 1, exhibits Epsilon Bootis with a power of three hundred and fifty; No. 2, Rigel, with a power of one hundred and thirty; No. 3, the Pole-star, with a power of one hundred; and No. 4, Castor, with a power of three hundred.

Our knowledge of the double stars almost commenced with Sir William Herschel, about the year 1780. At the time he began his search for them, he was acquainted with only *four*. Within five years he discovered nearly *seven hundred* double stars, and during his life, he observed no less than twenty-four hundred. In his Memoirs, published in the Philosophical Transactions, he gave most accurate measurements of the distances between the two stars, and of the angle which a line joining the two formed with a circle parallel to the equator. These data would enable him, or at least posterity, to judge whether these minute bodies ever change their position with respect to each other. Since 1821, these researches have been prosecuted, with great zeal and industry, by Sir James South and Sir John Herschel, in England; while Professor Struve, of Dorpat, with the celebrated telescope of Fraunhofer, has published, from his own observations, a catalogue of three thousand double stars, the determination of which involved the distinct and most minute inspection of at least one hundred and twenty thousand stars. Sir John Herschel, in his recent survey of the southern hemisphere, is said to have added to the catalogue of double stars nearly three thousand more.

Fig. 76.

Two circumstances add a high degree of interest to the phenomena of double stars: the first is, that a few of them, at least, are found to have a revolution around each other; the second, that they are supposed to afford the means of ascertaining the parallax of the fixed stars. But I must defer these topics till my next Letter.

LETTER XXIX.

FIXED STARS CONTINUED.

"O how canst thou renounce the boundless store
Of charms that Nature to her votary yields?
The warbling woodland, the resounding shore,
The pomp of groves, and garniture of fields;
All that the genial ray of morning yields,
 And all that echoes to the song of even,
All that the mountain's sheltering bosom shields,
And all the dread magnificence of heaven,—
O how canst thou renounce, and hope to be forgiven!"—*Beattie*.

IN 1803, Sir William Herschel first determined and announced to the world, that there exist among the stars separate systems, composed of two stars revolving about each other in regular orbits. These he denominated *binary stars*, to distinguish them from other double stars where no such motion is detected, and whose proximity to each other may possibly arise from casual juxtaposition, or from one being in the range of the other. Between fifty and sixty instances of changes, to a greater or less amount, of the relative positions of double stars, are mentioned by Sir William Herschel; and a few of them had changed their places so much, within twenty-five years, and in such order, as to lead him to the conclusion that they performed revolutions, one around the other, in regular orbits. These conclusions have been fully confirmed by later observers; so that it is now considered as fully established, that there exist among the fixed stars binary systems, in which two stars perform to each other the office of sun and planet, and that the periods of revolution of more than one such pair have been ascertained with some degree of exactness. Immersions and emersions of stars behind each other have been observed, and real motions among them detected, rapid enough to become sensible and measurable in very short intervals of time. The periods of the double stars are very various, ranging, in the case of those already ascertained, from forty-three years to one thousand. Their orbits are very small ellipses, only a few seconds in the longest direction, and more eccentric than those of the planets. A double star in the Northern Crown (*Eta Coronæ*) has made a complete revolution since its first discovery, and is now far advanced in its second period; while a star in the Lion (*Gamma Leonis*) requires twelve hundred years to complete its circuit.

You may not at once see the reason why these revolutions of one member of a double star around the other, should be deemed facts of such extraordinary interest; to you they may appear rather in the light of astronomical curiosities. But remark, that the revolutions of the binary stars have assured us of this most interesting fact, that *the law of gravitation extends to the fixed stars*. Before these discoveries, we could not decide, except by a feeble analogy, that this law transcended the bounds of the solar system. Indeed, our belief of the fact rested more upon our idea of unity of design in the works of the Creator, than upon any certain proof; but the revolution of one star around another, in obedience to forces which are proved to be similar to those which govern the solar system, establishes the grand conclusion, that the law of gravitation is truly the law of the material universe. "We have the same evidence," says Sir John Herschel, "of the revolutions of the binary stars about each other, that we have of those of Saturn and Uranus about the sun; and the correspondence between their calculated and observed places, in such elongated ellipses, must be admitted to carry with it a proof of the prevalence of the Newtonian law of gravity in their systems, of the very same nature and cogency as that of the calculated and observed places of comets round the centre of our own system. But it is not with the revolution of bodies of a cometary or planetary nature round a solar centre, that we are now concerned; it is with that of sun around sun, each, perhaps, accompanied with its train of planets and their satellites, closely shrouded from our view by the splendor of their respective suns, and crowded into a space, bearing hardly a greater proportion to the enormous interval which separates them, than the distances of the satellites of our planets from their primaries bear to their distances from the sun itself."

Many of the double stars are of different colors; and Sir John Herschel is of the opinion that there exist in nature suns of different colors. "It may," says he, "be easier suggested in words than conceived in imagination, what variety of illumination two suns, a red and a green, or a yellow and a blue one, must afford to a planet circulating about either; and what charming contrasts and 'grateful vicissitudes' a red and a green day, for instance, alternating with a white one and with darkness, might arise from the presence or absence of one or other or both above the horizon. Insulated stars of a red color, almost as deep as that of blood, occur in many parts of the heavens; but no green or blue star, of any decided hue, has ever been noticed unassociated with a companion brighter than itself."

Beside these revolutions of the binary stars, *some of the fixed stars appear to have a real motion in space*. There are several *apparent* changes of place among the stars, arising from real changes in the earth, which, as we are not conscious of them, we refer to the stars; but there are other motions among the stars which cannot result from any changes in the earth, but must arise from changes in the stars themselves. Such motions are called the *proper motions* of the stars. Nearly two thousand years ago, Hipparchus and Ptolemy made the most accurate determinations in their power of the relative situations of the stars, and their observations have been transmitted to us in Ptolemy's 'Almagest;' from which it appears that the stars retain at least *very nearly* the same places now as they did at that period. Still, the more accurate methods of modern astronomers have brought to light minute changes in the places of certain stars, which force upon us the conclusion, *either that our solar system causes an apparent displacement of certain stars, by a motion of its own in space, or that they have themselves a proper motion.* Possibly, indeed, both these causes may operate.

If the sun, and of course the earth which accompanies him, is actually in motion, the fact may become manifest from the apparent approach of the stars in the region which he is leaving, and the recession of those which lie in the part of the heavens towards which he is travelling. Were two groves of trees situated on a plain at some distance apart, and we should go from one to the other, the trees before us would gradually appear further and further asunder, while those we left behind would appear to approach each other. Some years since, Sir William Herschel supposed he had detected changes of this kind among two sets of stars in opposite points of the heavens, and announced that the solar system was in motion towards a point in the constellation Hercules; but other astronomers have not found the changes in question such as would correspond to this motion, or to any motion of the sun; and, while it is a matter of general belief that the sun has a motion in space, the fact is not considered as yet entirely proved.

In most cases, where a proper motion in certain stars has been suspected, its annual amount has been so small, that many years are required to assure us, that the effect is not owing to some other cause than a real progressive motion in the stars themselves; but in a few instances the fact is too obvious to admit of any doubt. Thus, the two stars, 61 Cygni, which are nearly equal, have remained constantly at the same or nearly at the same distance of fifteen seconds, for at least fifty years past. Meanwhile, they have shifted their local situation in the heavens four minutes twenty-three seconds, the annual proper motion of each star being five seconds and three tenths, by which quantity this system is every year carried along in some unknown path, by a motion which for many centuries must be regarded as uniform and rectilinear. A greater proportion of the double stars than of any other indicate proper motions, especially the binary stars, or those which have a revolution around each other. Among stars not double, and no way differing from the rest in any other obvious particular, a star in the constellation Cassiopeia, (*Mu Cassiopeiæ*) has the greatest proper motion of any yet ascertained, amounting to nearly four seconds annually.

You have doubtless heard much respecting the "immeasurable *distances*" of the fixed stars, and will desire to learn what is known to astronomers respecting this interesting subject.

We cannot ascertain the actual distance of any of the fixed stars, but we can certainly determine that the nearest star is more than twenty millions of millions of miles from the earth, (20,000,000,000,000.) For all measurements relating to the distances of the *sun and planets*, the radius of the earth furnishes the base line. The length of this line being known, and the horizontal parallax of the sun or any planet, we have the means of calculating the distance of the body from us, by methods explained in a previous Letter. But any star, viewed from the opposite sides of the earth, would appear from both stations to occupy precisely the same situation in the celestial sphere, and of course it would exhibit no horizontal parallax. But astronomers have endeavored to find a parallax in some of the fixed stars, by taking the *diameter of the earth's orbit* as a base line. Yet even a change of position amounting to one hundred and ninety millions of miles proved, until very recently, insufficient to alter the place of a single star, so far as to be capable of detection by very refined observations; from which it was concluded that the stars have not even any *annual parallax*; that is, the angle subtended by the semidiameter of the earth's orbit, at the nearest fixed star, is insensible. The errors to which instrumental measurements are subject, arising from the defects of instruments themselves, from refraction, and from various other sources of inaccuracy, are such, that the angular determinations of arcs of the heavens cannot be relied on to less than one second, and therefore cannot be appreciated by direct measurement. It follows, that, when viewed from the nearest star, the diameter of the earth's orbit would be insensible; the spider-line of the telescope would more than cover it. Taking, however, the annual parallax of a fixed star at one second, it can be demonstrated, that the distance of the nearest fixed star *must exceed* $95000000 \times 200000 = 190000000 \times 100000$, or one hundred thousand times one hundred and ninety millions of miles. Of a distance so vast we can form no adequate conceptions, and even seek to measure it only by the time that light (which moves more than one hundred and ninety-two thousand miles per second, and passes from the sun to the earth in eight minutes and seven seconds) would take to traverse it, which is found to be more than three and a half years.

If these conclusions are drawn with respect to the largest of the fixed stars, which we suppose to be vastly nearer to us than those of the smallest magnitude, the idea of distance swells upon us when we attempt to estimate the remoteness of the latter. As it is uncertain, however, whether the difference in the apparent magnitudes of the stars is owing to a real difference, or merely to their being at various distances from the eye, more or less uncertainty must attend all efforts to determine the relative distances of the stars; but astronomers generally believe, that the lower orders of stars are vastly more distant from us than the higher. Of some stars it is said, that thousands of years would be required for their light to travel down to us.

I have said that the stars have always been held, until recently, to have no annual parallax; yet it may be observed that astronomers were not exactly agreed on this point. Dr. Brinkley, a late eminent Irish astronomer, supposed that he had detected an annual parallax in Alpha Lyræ, amounting to one second and thirteen hundreths, and in Alpha Aquilæ, of one second and forty-two hundreths. These results were controverted by Mr. Pond, of the Royal Observatory of Greenwich; and Mr. Struve, of Dorpat, has shown that, in a number of cases, the supposed parallax is in a direction opposite to that which would arise from the motion of the earth. Hence it is considered doubtful whether, in all cases of an apparent parallax, the effect is not wholly due to errors of observation.

But as if nothing was to be hidden from our times, the long sought for parallax among the fixed stars has at length been found, and consequently the distance of some of these bodies, at least, is no longer veiled in mystery. In the year 1838, Professor Bessel, of Köningsberg, announced the discovery of a parallax in one of the stars of the Swan, (61 *Cygni*,) amounting to about *one third of a second.* This seems, indeed, so small an angle, that we might have reason to suspect the reality of the determination; but the most competent judges who have thoroughly examined the process by which the discovery was made, assent to its validity. What, then, do astronomers understand, when they say that a parallax has been discovered in one of the fixed stars, amounting to one third of a second? They mean that the star in question apparently shifts its place in the heavens, to that amount, when viewed at opposite extremities of the earth's orbit, namely, at points in space distant from each other one hundred and ninety millions of miles. On calculating the distance of the star from us from these data, it is found to be six hundred and fifty-seven thousand seven hundred times ninety-five millions of miles,—a distance which it would take light more than ten years to traverse.

Indirect methods have been proposed, for ascertaining the parallax of the fixed stars, by means of observations on the *double stars.* If the two stars composing a double star are at different distances from us, parallax would affect them unequally, and change their relative positions with respect to each other; and since the ordinary sources of error arising from the imperfection of instruments, from precession, and from refraction, would be avoided, (as they would affect both objects alike, and therefore would not disturb their relative positions,) measurements taken with the micrometer of changes much less than one second may be relied on. Sir John Herschel proposed a method, by which changes may be determined that amount to only one fortieth of a second.

The immense distance of the fixed stars is inferred also from the fact, that the largest telescopes do not increase their apparent magnitude. They are still points, when viewed with glasses that magnify five thousand times.

With respect to the NATURE OF THE STARS, it would seem fruitless to inquire into the nature of bodies so distant, and which reveal themselves to us only as shining points in space. Still, there are a few very satisfactory inferences that can be made out respecting them. First, *the fixed stars are bodies greater than our earth.* If this were not the case, they would not be visible at such an immense distance. Dr. Wollaston, a distinguished English philosopher, attempted to estimate the magnitudes of certain of the fixed stars from the light which they afford. By means of an accurate photometer, (an instrument for measuring the relative intensities of light,) he compared the light of Sirius with that of the sun. He next inquired how far the sun must be removed from us, in order to appear no brighter than Sirius. He found the distance to be one hundred and forty-one thousand times its present distance. But Sirius is more than two hundred thousand times as far off as the sun; hence he inferred that, upon the lowest computation, it must actually give out twice as much light as the sun; or that, in point of splendor, Sirius must be at least equal to two suns. Indeed, he has rendered it probable, that its light is equal to that of fourteen suns. There is reason, however, to believe that the stars are actually of various magnitudes, and that their apparent difference is not owing merely to their different distances. Bessel estimates the quantity of matter in the two members of a double star in the Swan, as less than half that of the sun.

Secondly, *the fixed stars are suns.* We have already seen that they are large bodies; that they are immensely further off than the furthest planet; that they shine by their own light; in short, that their appearance is, in all respects, the same as the sun would exhibit if removed to the region of the stars. Hence we infer that they are bodies of the same kind with the sun. We are justified, therefore, by a sound analogy, in concluding that the stars were made for the same end as the sun, namely, as the centres of attraction to other planetary worlds, to which they severally dispense light and heat. Although the starry heavens present, in a clear night, a spectacle of unrivalled grandeur and beauty, yet it must be admitted that the chief purpose of the stars could not have been to adorn the night, since by far the greater part of them are invisible to the naked eye; nor as landmarks to the navigator, for only a very small proportion of them are adapted to this purpose; nor, finally, to influence the earth by their attractions, since their distance renders such an effect entirely insensible. If they are suns, and if they exert no important agencies upon our world, but are bodies evidently adapted to the same purpose as our sun, then it is as rational to suppose that they were made to give light and heat, as that the eye was made for seeing and the ear for hearing. It is obvious to inquire, next, to what they dispense these gifts, if not to planetary worlds; and why to planetary worlds, if not for the use of percipient beings? We are thus led, almost inevitably, to the idea of a *plurality of worlds*; and the conclusion is forced upon us, that the spot which the Creator has assigned to us is but a humble province in his boundless empire.

LETTER XXX.

SYSTEM OF THE WORLD

"O how unlike the complex works of man,
Heaven's easy, artless, unincumbered, plan."—*Cowper.*

HAVING now explained to you, as far as I am able to do it in so short a space, the leading phenomena of the heavenly bodies, it only remains to inform you of the different systems of the world which have prevailed in different ages,—a subject which will necessarily involve a sketch of the history of astronomy.

By a system of the world, I understand an explanation of *the arrangement of all the bodies that compose the material universe, and of their relations to each other.* It is otherwise called the 'Mechanism of the Heavens;' and indeed, in the system of the world, we figure to ourselves a machine, all parts of which have a mutual dependence, and conspire to one great end. "The machines that were first invented," says Adam Smith, "to perform any particular movement, are always the most complex; and succeeding artists generally discover that, with fewer wheels, and with fewer principles of motion, than had originally been employed, the same effects may be more easily produced. The first systems, in the same manner, are always the most complex; and a particular connecting chain or principle is generally thought necessary, to unite every two seemingly disjointed appearances; but it often happens, that *one great connecting principle* is afterwards found to be sufficient to bind together all the discordant phenomena that occur in a whole species of things!" This remark is strikingly applicable to the origin and progress of systems of astronomy. It is a remarkable fact in the history of the human mind, that astronomy is the oldest of the sciences, having been cultivated, with no small success, long before any attention was paid to the causes of the common terrestrial phenomena. The opinion has always prevailed among those who were unenlightened by science, that very extraordinary appearances in the sky, as comets, fiery meteors, and eclipses, are omens of the wrath of heaven. They have, therefore, in all ages, been watched with the greatest attention: and their appearances have been minutely recorded by the historians of the times. The idea, moreover, that the aspects of the stars are connected with the destinies of individuals and of empires, has been remarkably prevalent from the earliest records of history down to a very late period, and, indeed, still lingers among the uneducated and credulous. This notion gave rise to ASTROLOGY,—an art which professed to be able, by a knowledge of the varying aspects of the planets and stars, to penetrate the veil of futurity, and to foretel approaching irregularities of Nature herself, and the fortunes of kingdoms and of individuals. That department of astrology which took cognizance of extraordinary occurrences in the natural world, as tempests, earthquakes, eclipses, and volcanoes, both to predict their approach and to interpret their meaning, was called *natural astrology*: that which related to the fortunes of men and of empires, *judicial astrology*. Among many ancient nations, astrologers were held in the highest estimation, and were kept near the persons of monarchs; and the practice of the art constituted a lucrative profession throughout the middle ages. Nor were the ignorant and uneducated portions of society alone the dupes of its pretensions. Hippocrates, the 'Father of Medicine,' ranks astrology among the most important branches of knowledge to the physician; and Tycho Brahe, and Lord Bacon, were firm believers in its mysteries. Astrology, fallacious as it was, must be acknowledged to have rendered the greatest services to astronomy, by leading to the accurate observation and diligent study of the stars.

At a period of very remote antiquity, astronomy was cultivated in China, India, Chaldea, and Egypt. The Chaldeans were particularly distinguished for the accuracy and extent of their astronomical observations. Calisthenes, the Greek philosopher who accompanied Alexander the Great in his Eastern conquests, transmitted to Aristotle a series of observations made at Babylon nineteen centuries before the capture of that city by Alexander; and the wise men of Babylon and the Chaldean astrologers are referred to in the Sacred Writings. They enjoyed a clear sky and a mild climate, and their pursuits as shepherds favored long-continued observations; while the admiration and respect accorded to the profession, rendered it an object of still higher ambition.

In the seventh century before the Christian era, astronomy began to be cultivated in Greece; and there arose successively three celebrated astronomical schools,—the school of Miletus, the school of Crotona, and the school of Alexandria. The first was established by Thales, six hundred and forty years before Christ; the second, by Pythagoras, one hundred and forty years afterwards; and the third, by the Ptolemies of Egypt, about three hundred years before the Christian era. As Egypt and Babylon were renowned among the most ancient nations, for their knowledge of the sciences, long before they were cultivated in Greece, it was the practice of the Greeks, when they aspired to the character of philosophers and sages, to resort to these countries to imbibe wisdom at its fountains. Thales, after extensive travels in Crete and Egypt, returned to his native place, Miletus, a town on the coast of Asia Minor, where he established the first school of astronomy in Greece. Although the minds of these ancient astronomers were beclouded with much error, yet Thales taught a few truths which do honor to his sagacity. He held that the stars are formed of fire; that the moon receives her light from the sun, and is invisible at her conjunctions because she is hid in the sun's rays. He taught the sphericity of the earth, but adopted the common error of placing it in the centre of the world. He introduced the division of the sphere into five zones, and taught the obliquity of the ecliptic. He was acquainted with the Saros, or sacred period of the Chaldeans, (see page 192,) and employed it in calculating eclipses. It was Thales that predicted the famous eclipse of the sun which terminated the war between the Lydians and the Medes, as mentioned in a former Letter. Indeed, Thales is universally regarded as a bright but solitary star, glimmering through mists on the distant horizon.

To Thales succeeded, in the school of Miletus, two other astronomers of much celebrity, Anaximander and Anaxagoras. Among many absurd things held by Anaximander, he first taught the sublime doctrine that the planets are inhabited, and that the stars are suns of other systems. Anaxagoras attempted to explain all the secrets of the skies by natural causes. His reasonings, indeed, were alloyed with many absurd notions; but still he alone, among the astronomers, maintained the existence of one God. His doctrines alarmed his countrymen, by their audacity and impiety to their gods, whose prerogatives he was thought to invade; and, to deprecate their wrath, sentence of death was pronounced on the philosopher and all his family,—a sentence which was commuted only for the sad alternative of perpetual banishment. The very genius of the heathen mythology was at war with the truth. False in itself, it trained the mind to the love of what was false in the interpretation of nature; it arrayed itself against the simplicity of truth, and persecuted and put to death its most ardent votaries. The religion of the Bible, on the other hand, lends all its aid to truth in nature as well as in morals and religion. In its very genius it inculcates and inspires the love of truth; it suggests, by its analogies, the existence of established laws in the system of the world; and holds out the moon and the stars, which the Creator has ordained, as fit objects to give us exalted views of his glory and wisdom.

Pythagoras was the founder of the celebrated school of Crotona. He was a native of Samos, an island in the Ægean sea, and flourished about five hundred years before the Christian era. After travelling more than thirty years in Egypt and Chaldea, and spending several years more at Sparta, to learn the laws and institutions of Lycurgus, he returned to his native island to dispense the riches he had acquired to his countrymen. But they, probably fearful of incurring the displeasure of the gods by the freedom with which he inquired into the secrets of the skies, gave him so unwelcome a reception, that he retired from them, in disgust, and established his school at Crotona, on the southeastern coast of Italy.

Hither, as to an oracle, the fame of his wisdom attracted hundreds of admiring pupils, whom he instructed in every species of knowledge. From the visionary notions which are generally understood to have been entertained on the subject of astronomy, by the ancients, we are apt to imagine that they knew less than they actually did of the truths of this science. But Pythagoras was acquainted with many important facts in astronomy, and entertained many opinions respecting the system of the world, which are now held to be true. Among other things well known to Pythagoras, either derived from his own investigations, or received from his predecessors, were the following; and we may note them as a synopsis of the state of astronomical knowledge at that age of the world. First, the principal *constellations*. These had begun to be formed in the earliest ages of the world. Several of them, bearing the same name as at present, are mentioned in the writings of Hesiod and Homer; and the "sweet influences of the Pleiades," and the "bands of Orion," are beautifully alluded to in the book of Job. Secondly, *eclipses*. Pythagoras knew both the causes of eclipses and how to predict them; not, indeed, in the accurate manner now practised, but by means of the Saros. Thirdly, Pythagoras had divined the true *system of the world*, holding that the sun, and not the earth, (as was generally held by the ancients, even for many ages after Pythagoras,) is the centre around which all the planets revolve; and that the stars are so many suns, each the centre of a system like our own. Among lesser things, he knew that the earth is round; that its surface is naturally divided into five zones; and that the ecliptic is inclined to the equator. He also held that the earth revolves daily on its axis, and yearly around the sun; that the galaxy is an assemblage of small stars; and that it is the same luminary, namely, Venus, that constitutes both the morning and evening star; whereas all the ancients before him had supposed that each was a separate planet, and accordingly the morning star was called Lucifer, and the evening star, Hesperus. He held, also, that the planets were inhabited, and even went so far as to calculate the size of some of the animals in the moon. Pythagoras was also so great an enthusiast in music, that he not only assigned to it a conspicuous place in his system of education, but even supposed that the heavenly bodies themselves were arranged at distances corresponding to the intervals of the diatonic scale, and imagined them to pursue their sublime march to notes created by their own harmonious movements, called the 'music of the spheres;' but he maintained that this celestial concert, though loud and grand, is not audible to the feeble organs of man, but only to the gods. With few exceptions, however, the opinions of Pythagoras on the system of the world were founded in truth. Yet they were rejected by Aristotle, and by most succeeding astronomers, down to the time of Copernicus; and in their place was substituted the doctrine of *crystalline spheres*, first taught by Eudoxus, who lived about three hundred and seventy years before Christ. According to this system, the heavenly bodies are set like gems in hollow solid orbs, composed of crystal so transparent, that no anterior orb obstructs in the least the view of any of the orbs that lie behind it. The sun and the planets have each its separate orb; but the fixed stars are all set in the same grand orb; and beyond this is another still, the *primum mobile*, which revolves daily, from east to west, and carries along with it all the other orbs. Above the whole spreads the *grand empyrean*, or third heavens, the abode of perpetual serenity.

To account for the planetary motions, it was supposed that each of the planetary orbs, as well as that of the sun, has a motion of its own, eastward, while it partakes of the common diurnal motion of the starry sphere. Aristotle taught that these motions are effected by a tutelary genius of each planet, residing in it, and directing its motions, as the mind of man directs his movements.

Two hundred years after Pythagoras, arose the famous school of Alexandria, under the Ptolemies. These were a succession of Egyptian kings, and are not to be confounded with Ptolemy, the astronomer. By the munificent patronage of this enlightened family, for the space of three hundred years, beginning at the death of Alexander the Great, from whom the eldest of the Ptolemies had received his kingdom, the school of Alexandria concentrated in its vast library and princely halls, erected for the accommodation of the philosophers, nearly all the science and learning of the world. In wandering over the immense territories of ignorance and barbarism which covered, at that time, almost the entire face of the earth, the eye reposes upon this little spot, as upon a verdant island in the midst of the desert. Among the choice fruits that grew in this garden of astronomy were several of the most distinguished ornaments of ancient science, of whom the most eminent were Hipparchus and Ptolemy. Hipparchus is justly considered as the Newton of antiquity. He sought his knowledge of the heavenly bodies not in the illusory suggestions of a fervid imagination, but in the vigorous application of an intellect of the first order. Previous to this period, celestial observations were made chiefly with the naked eye: but Hipparchus was in possession of instruments for measuring angles, and knew how to resolve spherical triangles. These were great steps beyond all his predecessors. He ascertained the length of the year within six minutes of the truth. He discovered the eccentricity, or elliptical figure, of the solar orbit, although he supposed the sun actually to move uniformly in a circle, but the earth to be placed out of the centre. He also determined the positions of the points among the stars where the earth is nearest to the sun, and where it is most remote from it. He formed very accurate estimates of the obliquity of the ecliptic and of the precession of the equinoxes. He computed the exact period of the synodic revolution of the moon, and the inclination of the lunar orbit; discovered the backward motion of her node and of her line of apsides; and made the first attempts to ascertain the horizontal parallaxes of the sun and moon. Upon the appearance of a new star in the firmament, he undertook, as already mentioned, to number the stars, and to assign to each its true place in the heavens, in order that posterity might have the means of judging what changes, if any, were going forward among these apparently unalterable bodies.

Although Hipparchus is generally considered as belonging to the Alexandrian school, yet he lived at Rhodes, and there made his astronomical observations, about one hundred and forty years before the Christian era. One of his treatises has come down to us; but his principal discoveries have been transmitted through the 'Almagest' of Ptolemy. Ptolemy flourished at Alexandria nearly three centuries after Hipparchus, in the second century after Christ. His great work, the 'Almagest,' which has conveyed to us most that we know respecting the astronomical knowledge of the ancients, was the universal text-book of astronomers for fourteen centuries.

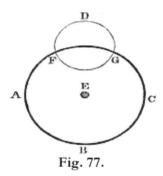

Fig. 77.

The name of this celebrated astronomer has also descended to us, associated with the system of the world which prevailed from Ptolemy to Copernicus, called the *Ptolemaic System*. The doctrines of the Ptolemaic system did not originate with Ptolemy, but, being digested by him out of materials furnished by various hands, it has come down to us under the sanction of his name. According to this system, the earth is the centre of the universe, and all the heavenly bodies daily revolve around it, from east to west. But although this hypothesis would account for the apparent diurnal motion of the firmament, yet it would not account for the apparent annual motion of the sun, nor for the slow motions of the planets from west to east. In order to explain these phenomena, recourse was had to *deferents* and *epicycles*,—an explanation devised by Apollonius, one of the greatest geometers of antiquity. He conceived that, in the circumference of a circle, having the earth for its centre, there moves the centre of a smaller circle in the circumference of which the planet revolves. The circle surrounding the earth was called the deferent, while the smaller circle, whose centre was always in the circumference of the deferent, was called the epicycle. Thus, if E, Fig. 77, represents the earth, ABC will be the deferent, and DFG, the epicycle; and it is obvious that the motion of a body from west to east, in this small circle, would be alternately direct, stationary, and retrograde, as was explained, in a previous Letter, to be actually the case with the apparent motions of the planets. The hypothesis, however, is inconsistent with the *phases* of Mercury and Venus, which, being between us and the sun, on both sides of the epicycle, would present their dark sides towards us at both conjunctions with the sun, whereas, at one of the conjunctions, it is known that they exhibit their disks illuminated. It is, moreover, absurd to speak of a geometrical centre, which has no bodily existence, moving round the earth on the circumference of another circle. In addition to these absurdities, the whole Ptolemaic system is encumbered with the following difficulties: First, it is a mere hypothesis, having no evidence in its favor except that it explains the phenomena. This evidence is insufficient of itself, since it frequently happens that each of two hypotheses, which are directly opposite to each other, will explain all the known phenomena. But the Ptolemaic system does not even do this, as it is inconsistent with the phases of Mercury and Venus, as already observed. Secondly, now that we are acquainted with the distances of the remoter planets, and especially the fixed stars, the swiftness of motion, implied in a daily revolution of the starry firmament around the earth, renders such a motion wholly incredible. Thirdly, the centrifugal force which would be generated in these bodies, especially in the sun, renders it impossible that they can continue to revolve around the earth as a centre. Absurd, however, as the system of Ptolemy was, for many centuries no great philosophic genius appeared to expose its fallacies, and it therefore guided the faith of astronomers of all countries down to the time of Copernicus.

After the age of Ptolemy, the science made little progress. With the decline of Grecian liberty, the arts and sciences declined also; and the Romans, then masters of the world, were ever more ambitious to gain conquests over man than over matter; and they accordingly never produced a single great astronomer. During the middle ages, the Arabians were almost the only astronomers, and they cultivated this noble study chiefly as subsidiary to astrology.

At length, in the fifteenth century, Copernicus arose, and after forty years of intense study and meditation, divined the true system of the world. You will recollect that the Copernican system maintains, 1. That the *apparent* diurnal motions of the heavenly bodies, from east to west, is owing to the *real* revolution of the earth on its own axis from west to east; and, 2. That the sun is the centre around which the earth and planets all revolve from west to east. It rests on the following arguments: In the first place, *the earth revolves on its own axis.* First, because this supposition is vastly more *simple.* Secondly, it is agreeable to *analogy*, since all the other planets that afford any means of determining the question, are seen to revolve on their axes. Thirdly, the *spheroidal figure* of the earth is the figure of equilibrium, that results from a revolution on its axis. Fourthly, the *diminished weight* of bodies at the equator indicates a centrifugal force arising from such a revolution. Fifthly, bodies let fall from a high eminence, fall *eastward of their base*, indicating that when further from the centre of the earth they were subject to a greater velocity, which, in consequence of their inertia, they do not entirely lose in descending to the lower level.

In the second place, *the planets, including the earth, revolve about the sun.* First, the *phases* of Mercury and Venus are precisely such, as would result from their circulating around the sun in orbits within that of the earth; but they are never seen in opposition, as they would be, if they circulate around the earth. Secondly, the superior planets do indeed revolve around the earth; but they also revolve around the sun, as is evident from their phases, and from the known dimensions of their orbits; and that the sun, and not the earth, is the *centre* of their motions, is inferred from the greater symmetry of their motions, as referred to the sun, than as referred to the earth; and especially from the laws of gravitation, which forbid our supposing that bodies so much larger than the earth, as some of these bodies are, can circulate permanently around the earth, the latter remaining all the while at rest.

In the third place, the annual motion of *the earth* itself is indicated also by the most conclusive arguments. For, first, since all the planets, with their satellites and the comets, revolve about the sun, analogy leads us to infer the same respecting the earth and its satellite, as those of Jupiter and Saturn, and indicates that it is a law of the solar system that the smaller bodies revolve about the larger. Secondly, on the supposition that the earth performs an annual revolution around the sun, it is embraced along with the planets, in Kepler's law, that the squares of the times are as the cubes of the distances; otherwise, it forms an exception, and the only known exception, to this law.

Such are the leading arguments upon which rests the Copernican system of astronomy. They were, however, only very partially known to Copernicus himself, as the state both of mechanical science, and of astronomical observation, was not then sufficiently matured to show him the strength of his own doctrine, since he knew nothing of the telescope, and nothing of the principle of universal gravitation. The evidence of this beautiful system being left by Copernicus in so imperfect a state, and indeed his own reasonings in support of it being tinctured with some errors, we need not so much wonder that Tycho Brahe, who immediately followed Copernicus, did not give it his assent, but, influenced by certain passages of Scripture, he still maintained, with Ptolemy, that the earth is in the centre of the universe; and he accounted for the diurnal motions in the same manner as Ptolemy had done, namely, by an actual revolution of the whole host of heaven around the earth every twenty-four hours. But he rejected the scheme of deferents and epicycles, and held that the moon revolves about the earth as the centre of her motions; but that the sun and not the earth is the centre of the planetary motions; and that the sun, accompanied by the planets, moves around the earth once a year, somewhat in the manner in which we now conceive of Jupiter and his satellites as revolving around the sun. This system is liable to most of the objections that lie against the Ptolemaic system, with the disadvantage of being more complex.

Kepler and Galileo, however, as appeared in the sketch of their lives, embraced the theory of Copernicus with great avidity, and all their labors contributed to swell the evidence of its truth. When we see with what immense labor and difficulty the disciples of Ptolemy sought to reconcile every new phenomenon of the heavens with their system, and then see how easily and naturally all the successive discoveries of Galileo and Kepler fall in with the theory of Copernicus, we feel the full force of those beautiful lines of Cowper which I have chosen for the motto of this Letter.

Newton received the torch of truth from Galileo, and transmitted it to his successors, with its light enlarged and purified; and since that period, every new discovery, whether the fruit of refined instrumental observation or of profound mathematical analysis, has only added lustre to the glory of Copernicus.

With Newton commenced a new and wonderful era in astronomy, distinguished above all others, not merely for the production of the greatest of men, but also for the establishment of those most important auxiliaries to our science, the Royal Society of London, the Academy of Sciences at Paris, and the Observatory of Greenwich. I may add the commencement of the Transactions of the Royal Society, and the Memoirs of the Academy of Sciences, which have been continued to the present time,—both precious storehouses of astronomical riches. The Observatory of Greenwich, moreover, has been under the direction of an extraordinary succession of great astronomers. Their names are Flamstead, Halley, Bradley, Maskeleyne,

Pond, and Airy,—the last being still at his post, and worthy of continuing a line so truly illustrious. The observations accumulated at this celebrated Observatory are so numerous, and so much superior to those of any other institution in the world, that it has been said that astronomy would suffer little, if all other contemporary observations of the same kind were annihilated. Sir William Herschel, however, labored chiefly in a different sphere. The Astronomers Royal devoted themselves not so much to the discovery of new objects among the heavenly bodies, as to the exact determination of the places of the bodies already known, and to the developement of new laws or facts among the celestial motions. But Herschel, having constructed telescopes of far greater reach than any ever used before, employed them to sound new and untried depths in the profundities of space. We have already seen what interesting and amazing discoveries he made of double stars, clusters, and nebulæ.

The English have done most for astronomy in observation and discovery; but the French and Germans, in developing, by the most profound mathematical investigation, the great laws of physical astronomy.

It only remains to inquire, whether the Copernican system is now to be regarded as a full exposition of the 'Mechanism of the Heavens,' or whether there subsist higher orders of relations between the fixed stars themselves.

The revolutions of the *binary stars* afford conclusive evidence of at least subordinate systems of suns, governed by the same laws as those which regulate the motions of the solar system. The *nebulæ* also compose peculiar systems, in which the members are evidently bound together by some common relation.

In these marks of organization,—of stars associated together in clusters; of sun revolving around sun; and of nebulæ disposed in regular figures,— we recognise different members of some grand system, links in one great chain that binds together all parts of the universe; as we see Jupiter and his satellites combined in one subordinate system, and Saturn and his satellites in another,—each a vast kingdom, and both uniting with a number of other individual parts, to compose an empire still more vast.

This fact being now established, that the stars are immense bodies, like the sun, and that they are subject to the laws of gravitation, we cannot conceive how they can be preserved from falling into final disorder and ruin, unless they move in harmonious concert, like the members of the solar system. Otherwise, those that are situated on the confines of creation, being retained by no forces from without, while they are subject to the attraction of all the bodies within, must leave their stations, and move inward with accelerated velocity; and thus all the bodies in the universe would at length fall together in the common centre of gravity. The immense distance at which the stars are placed from each other would indeed delay such a catastrophe; but this must be the ultimate tendency of the material world, unless sustained in one harmonious system by nicely-adjusted motions. To leave entirely out of view our confidence in the wisdom and preserving goodness of the Creator, and reasoning merely from what we know of the stability of the solar system, we should be justified in inferring, that other worlds are not subject to forces which operate only to hasten their decay, and to involve them in final ruin.

We conclude, therefore, that the material universe is one great system; that the combination of planets with their satellites constitutes the first or lowest order of worlds; that next to these, planets are linked to suns; that these are bound to other suns, composing a still higher order in the scale of being; and finally, that all the different systems of worlds move around their common centre of gravity.

LETTER XXXI.

NATURAL THEOLOGY.

——"Philosophy, baptized
In the pure fountain of Eternal Love,
Has eyes indeed; and, viewing all she sees
As meant to indicate a God to man,
Gives Him the praise, and forfeits not her own."—*Cowper.*

I INTENDED, my dear Friend, to comply with your request "that I would discuss the arguments which astronomy affords to natural theology;" but these Letters have been already extended so much further than I anticipated, that I shall conclude with suggesting a few of those moral and religious reflections, which ought always to follow in the train of such a survey of the heavenly bodies as we have now taken.

Although there is evidence enough in the structure, arrangement, and laws, which prevail among the heavenly bodies, to prove the *existence* of God, yet I think there are many subordinate parts of His works far better adapted to this purpose than these, being more fully within our comprehension. It was intended, no doubt, that the evidence of His being should be accessible to all His creatures, and should not depend on a kind of knowledge possessed by comparatively few. The mechanism of the eye is probably not more perfect than that of the universe; but we can analyze it better, and more fully understand the design of each part. But the existence of God being once proved, and it being admitted that He is the Creator and Governor of the world, then the discoveries of astronomy are admirably adapted to perform just that office in relation to the Great First Cause, which is assigned to them in the Bible, namely, "to declare the glory of God, and to show His handiwork." In other words, the discoveries of astronomy are peculiarly fitted,—more so, perhaps, than any other department of creation,—to exhibit the unity, power, and wisdom, of the Creator.

The most modern discoveries have multiplied the proofs of the *unity* of God. It has usually been offered as sufficient evidence of the truth of this doctrine, that the laws of Nature are found to be uniform when applied to the utmost bounds of the *solar system*; that the law of gravitation controls alike the motions of Mercury, and those of Uranus; and that its operation is one and the same upon the moon and upon the satellites of Saturn. It was, however, impossible, until recently, to predicate the same uniformity in the great laws of the universe respecting the starry worlds, except by a feeble analogy. However improbable, it was still possible, that in these distant worlds other laws might prevail, and other Lords exercise dominion. But the discovery of the revolutions of the binary stars, in exact accordance with the law of gravitation, not merely in a single instance, but in many instances, in all cases, indeed, wherever those revolutions have advanced so far as to determine their law of action, gives us demonstration, instead of analogy, of the prevalence of the same law among the other systems as that which rules in ours.

The marks of a still higher organization in the structure of clusters and nebulæ, all bearing that same characteristic union of resemblance and variety which belongs to all the other works of creation that fall under our notice, speak loudly of one, and only one, grand design. Every new discovery of the telescope, therefore, has added new proofs to the great truth that God is one: nor, so far as I know, has a single fact appeared, that is not entirely consonant with it. Light, moreover, which brings us intelligence, and, in most cases, the only intelligence we have, of these remote orbs, testifies to the same truth, being similar in its properties and uniform in its motions, from whatever star it emanates.

In displays of the *power* of Jehovah, nothing can compare with the starry heavens. The magnitudes, distances, and velocities, of the heavenly bodies are so much beyond every thing of this kind which belongs to things around us, from which we borrowed our first ideas of these qualities, that we can scarcely avoid looking with incredulity at the numerical results to which the unerring principles of mathematics have conducted us. And when we attempt to apply our measures to the fixed stars, and especially to the nebulæ, the result is absolutely overwhelming: the mind refuses its aid in our attempts to grasp the great ideas. Nor less conspicuous, among the phenomena of the heavenly bodies, is the *wisdom* of the Creator. In the first place, this attribute is every where exhibited *in the happy adaptation of means to their ends.* No principle can be imagined more simple, and at the same time more effectual to answer the purposes which it serves, than gravitation. No position can be given to the sun and planets so fitted, as far as we can judge, to fulfil their mutual relations, as that which the Creator has given them. I say, as far as we can judge; for we find this to be the case in respect to our own planet and its attendant satellite, and hence have reason to infer that the same is the case in the other planets, evidently holding, as they do, a similar relation to the sun. Thus the position of the earth at just such a distance from the sun as suits the nature of its animal and vegetable kingdoms, and confining the range of solar heat, vast as it might easily become, within such narrow bounds; the inclination of the earth's axis to the plane of its orbit, so as to produce the agreeable vicissitudes of the seasons, and increase the varieties of animal and vegetable life, still confining the degree of inclination so exactly within the bounds of safety, that, were it much to transcend its present limits, the changes of temperature of the different seasons would be too sudden and violent for the existence of either animals or vegetables; the revolution of the earth on its axis, so happily dividing time into hours of business and of repose; the adaptation of the moon to the earth, so as to afford to us her greatest amount of light just at the times when it is needed most, and giving to the moon just such a quantity of matter, and placing her at just such a distance from the earth, as serves to raise a tide productive of every conceivable advantage, without the evils which would result from a stagnation of the waters on the one hand, or from their overflow on the other;—these are a few examples of the wisdom displayed in the mutual relations instituted between the sun, the earth, and the moon.

In the second place, similar marks of wisdom are exhibited in *the many useful and important purposes which the same thing is made to serve*. Thus the sun is at once the great regulator of the planetary motions, and the fountain of light and heat. The moon both gives light by night and raises the tides. Or, if we would follow out this principle where its operations are more within our comprehension, we may instance the *atmosphere*. When man constructs an instrument, he deems it sufficient if it fulfils one single purpose as the watch, to tell the hour of the day, or the telescope, to enable him to see distant objects; and had a being like ourselves made the atmosphere, he would have thought it enough to have created a medium so essential to animal life, that to live is to breathe, and to cease to breathe is to die. But beside this, the atmosphere has manifold uses, each entirely distinct from all the others. It conveys to plants, as well as animals, their nourishment and life; it tempers the heat of Summer with its breezes; it binds down all fluids, and prevents their passing into the state of vapor; it supports the clouds, distils the dew, and waters the earth with showers; it multiplies the light of the sun, and diffuses it over earth and sky; it feeds our fires, turns our machines, wafts our ships, and conveys to the ear all the sentiments of language, and all the melodies of music.

In the third place, the wisdom of the Creator is strikingly manifested in the provision he has made for the *stability of the universe*. The perturbations occasioned by the motions of the planets, from their action on each other, are very numerous, since every body in the system exerts an attraction on every other, in conformity with the law of universal gravitation. Venus and Mercury, approaching, as they do at times, comparatively near to the earth, sensibly disturb its motions; and the satellites of the remoter planets greatly disturb each other's movements. Nor was it possible to endow this principle with the properties it has, and make it operate as it does in regulating the motions of the world, without involving such an incident. On this subject, Professor Whewell, in his excellent work composing one of the Bridgewater Treatises, remarks: "The derangement which the planets produce in the motion of one of their number will be very small, in the course of one revolution; but this gives us no security that the derangement may not become very large, in the course of many revolutions. The cause acts perpetually, and it has the whole extent of time to work in. Is it not easily conceivable, then, that, in the lapse of ages, the derangements of the motions of the planets may accumulate, the orbits may change their form, and their mutual distances may be much increased or diminished? Is it not possible that these changes may go on without limit, and end in the complete subversion and ruin of the system? If, for instance, the result of this mutual gravitation should be to increase considerably the eccentricity of the earth's orbit, or to make the moon approach continually nearer and nearer to the earth, at every revolution, it is easy to see that, in the one case, our year would change its character, producing a far greater irregularity in the distribution of the solar heat; in the other, our satellite must fall to the earth, occasioning a dreadful catastrophe. If the positions of the planetary orbits, with respect to that of the earth, were to change much, the planets might sometimes come very near us, and thus increase the effect of their attraction beyond calculable limits. Under such circumstances, 'we might have years of unequal length, and seasons of capricious temperature; planets and moons, of portentous size and aspect, glaring and disappearing at uncertain intervals; tides, like deluges, sweeping over whole continents; and perhaps the collision of two of the planets, and the consequent destruction of all organization on both of them.' The fact really is, that changes are taking place in the motions of the heavenly bodies, which have gone on progressively, from the first dawn of science. The eccentricity of the earth's orbit has been diminishing from the earliest observations to our times. The moon has been moving quicker from the time of the first recorded eclipses, and is now in advance, by about four times her own breadth, of what her own place would have been, if it had not been affected by this acceleration. The obliquity of the ecliptic, also, is in a state of diminution, and is now about two fifths of a degree less than it was in the time of Aristotle."

But amid so many seeming causes of irregularity and ruin, it is worthy of a grateful notice, that effectual provision is made for the *stability of the solar system*. The full confirmation of this fact is among the grand results of physical astronomy. "Newton did not undertake to demonstrate either the stability or instability of the system. The decision of this point required a great number of preparatory steps and simplifications, and such progress in the invention and improvement of mathematical methods, as occupied the best mathematicians of Europe for the greater part of the last century. Towards the end of that time, it was shown by La Grange and La Place, that the arrangements of the solar system are stable; that, in the long run, the orbits and motions remain unchanged; and that the changes in the orbits, which take place in shorter periods, never transgress certain very moderate limits. Each orbit undergoes deviations on this side and on that side of its average state; but these deviations are never very great, and it finally recovers from them, so that the average is preserved. The planets produce perpetual perturbations in each other's motions; but these perturbations are not indefinitely progressive, but periodical, reaching a maximum value, and then diminishing. The periods which this restoration requires are, for the most part, enormous,—not less than thousands, and in some instances, millions, of years. Indeed, some of these apparent derangements have been going on in the same direction from the creation of the world. But the restoration is in the sequel as complete as the derangement; and in the mean time the disturbance never attains a sufficient amount seriously to affect the stability of the system. 'I have succeeded in demonstrating,' says La Place, 'that, whatever be the masses of the planets, in consequence of the fact that they all move in the same direction, in orbits of small eccentricity, and but slightly inclined to each other, their secular irregularities are periodical, and included within narrow limits; so that the planetary system will only oscillate about a mean state, and will never deviate from it, except by a very small quantity. The ellipses of the planets have been and always will be nearly circular. The ecliptic will never coincide with the equator; and the entire extent of the variation, in its inclination, cannot exceed three degrees.'"

To these observations of La Place, Professor Whewell adds the following, on the importance, to the stability of the solar system, of the fact that those planets which have *great masses* have orbits of *small eccentricity*. "The planets Mercury and Mars, which have much the largest eccentricity among the old planets, are those of which the masses are much the smallest. The mass of Jupiter is more than two thousand times that of either of these planets. If the orbit of Jupiter were as eccentric as that of Mercury, all the security for the stability of the system, which analysis has yet pointed out, would disappear. The earth and the smaller planets might, by the near approach of Jupiter at his perihelion, change their nearly circular orbits into very long ellipses, and thus might fall into the sun, or fly off into remoter space. It is further remarkable, that in the newly-discovered planets, of which the orbits are still more eccentric than that of Mercury, the masses are still smaller, so that the same provision is established in this case, also."

With this hasty glance at the unity, power, and wisdom, of the Creator, as manifested in the greatest of His works, I close. I hope enough has been said to vindicate the sentiment that called 'Devotion, daughter of Astronomy!' I do not pretend that this, or any other science, is adequate of itself to purify the heart, or to raise it to its Maker; but I fully believe that, when the heart is already under the power of religion, there is something in the frequent and habitual contemplation of the heavenly bodies under all the lights of modern astronomy, very favorable to devotional feelings, inspiring, as it does, humility, in unison with an exalted sentiment of grateful adoration.

LETTER XXXII.

RECENT DISCOVERIES.

"All are but parts of one stupendous whole."—*Pope*.

WITHIN a few years, astronomy has been enriched with a number of valuable discoveries, of which I will endeavor to give you a summary account in this letter. The heavens have been explored with far more powerful telescopes than before; instrumental measurements have been carried to an astonishing degree of accuracy; numerous additions have been made to the list of small planets or asteroids; a comet has appeared of extraordinary splendor, remarkable, above all others, for its near approach to the sun; the distances of several of the fixed stars, an element long sought for in vain, have been determined; a large planet, composing in itself a magnificent world, has been added to the solar system, at such a distance from the central luminary as nearly to double the supposed dimensions of that system; various nebulæ, before held to be irresolvable, have been resolved into stars; and a new satellite has been added to Saturn.

IMPROVEMENTS IN THE TELESCOPE.—Herschel's forty-feet telescope, of which I gave an account in my fourth letter (see page 36), remained for half a century unequalled in magnitude and power; but in 1842, Lord Rosse, an Irish nobleman, commenced a telescope on a scale still more gigantic. Like Herschel's, it was a *reflector*, the image being formed by a concave mirror. This was six feet in diameter, and weighed three tons; and the tube was fifty feet in length. The entire cost of the instrument was sixty thousand dollars. Its reflecting surface is nearly twice as great as the great Herschelian, and consequently it greatly exceeds all instruments hitherto constructed in the *amount of light* which it collects and transmits to the eye; and this adapts it peculiarly to viewing those objects, as nebulæ, whose light is exceedingly faint. Accordingly, it has revealed to us new wonders in this curious department of astronomy. Some idea of the great dimensions of the *Leviathan* telescope (as this instrument has been called) may be formed when it is said that the Dean of Ely, a full-sized man, walked through the tube from one end to the other, with an umbrella over his head.

But still greater advances have been made in refracting than in reflecting telescopes. Such was the difficulty of obtaining large pieces of glass which are free from impurities, and such the liability of large lenses to form obscure and colored images, that it was formerly supposed impossible to make a refracting telescope larger in diameter than five or six inches; but their size has been increased from one step to another, until they are now made more than fifteen inches in diameter; and so completely have all the difficulties arising from the imperfections of glass, and from optical defects inherent in lenses, been surmounted, that the great telescopes of Pulkova, at St. Petersburgh, and of Harvard University (the two finest refractors in the world) are considered among the most perfect productions of the arts. A lens of only 15 inches in diameter seems, indeed, diminutive when compared with a concave reflector of six feet; but for most purposes of the astronomer, the Pulkova and Cambridge instruments are more useful than such great reflectors as those of Herschel and Rosse. If there is any particular in which these are more effective, it is in observations on the faintest nebulæ, where it is necessary to collect and convey to the eye the greatest possible beam of light.

INSTRUMENTAL MEASUREMENTS.—When astronomical instruments were first employed to measure the angular distance between two points on the celestial sphere, it was not attempted to measure spaces smaller than ten minutes—a space equal to the third part of the breadth of the full moon. Tycho Brahe, however, carried his measures to sixty times that degree of minuteness, having devised means of determining angles no larger than ten seconds, or the one hundred and eightieth part of the breadth of the lunar disk. For many years past, astronomers have carried these measures to single seconds, or have determined spaces no greater than the eighteen hundredth part of the diameter of the moon. This is considered the smallest arc which can be accurately measured directly on the limb of an instrument; but *differences* between spaces may be estimated to a far greater degree of accuracy than this, even to the hundredth part of a second—a space less than that intercepted by a spider's web held before the eye.

DISCOVERY OF NEW PLANETS.—In my twenty-third letter (see page 286), I gave an account of the small planets called asteroids, which lie between the orbits of Mars and Jupiter. When that letter was written, no longer ago than 1840, only four of those bodies had been discovered, namely, Ceres, Pallas, Juno, and Vesta. Within a few years past, nineteen more have been added, making the number of the asteroids known at present twenty-three, and every year adds one or more to the list.[17] The idea first suggested by Olbers, one of the earliest discoverers of asteroids, that they are fragments of a large single planet once revolving between Mars and Jupiter, has gained credit since the discovery of so many additional bodies of the same class, all, like the former, exceedingly small and irregular in their motions, although there are still great difficulties in tracing them to a common origin.

GREAT COMET OF 1843.—This is the most wonderful body that has appeared in the heavens in modern times; first, on account of its appearing, when first seen, in the broad light of noonday; and, secondly, on account of its approaching so near the sun as almost to graze his surface. It was first discovered, in New England, on the 28th of February, a little eastward of the sun, shining like a white cloud illuminated by the solar rays. It arrested the attention of many individuals from half past seven in the morning until three o'clock in the afternoon, when the sky became obscured by clouds. In Mexico, it was observed from nine in the morning until sunset. At a single station in South America, it was said to have been seen on the 27th of February, almost in contact with the sun. Early in March, it had receded so far to the eastward of that body as to be visible in the southwest after sunset, throwing upward a long train, which increased in length from night to night until it covered a space of 40 degrees. Its position may be seen on a celestial globe adjusted to the latitude of New Haven (41° 18´) for the 20th of March, by tracing a line, or, rather, a broad band proceeding from the place of the sun towards the bright star Sirius, in the south, between the ears of the Hare and the feet of Orion.

The comet passed its perihelion on the 27th of February, at which time it almost came in contact with the sun. To prevent its falling into the sun it was endued with a prodigious velocity; a velocity so great that, had it continued at the same rate as at the instant of perihelion passage, it would have whirled round the sun in two hours and a half. It did, in fact, complete more than half its revolution around the sun in that short period, and it made more than three quarters of its circuit around the sun in one day. Its velocity, when nearest the sun, exceeded a million of miles per hour, and its tail, at its greatest elongation, was one hundred and eight millions of miles; a length more than sufficient to have reached from the sun to the earth. Its heat was estimated to be 47,000 times greater than that received by the earth from a vertical sun, and consequently it was more intense than that produced by the most powerful blowpipes, and sufficient to melt like wax the most infusible bodies. No doubt, when in the vicinity of the sun, the solid matter of the comet was first melted and then converted into vapor, which itself became red hot, or, more properly speaking, *white hot*. Much discussion has arisen among astronomers respecting the periodic time of this comet. Its most probable period is about 175 years.

DISTANCES OF THE STARS.—I have already mentioned (page 389) that the distance of at least one of the fixed stars has at length been determined, although at so great a distance that its annual parallax is only about one third of a second, implying a distance from the sun of nearly sixty millions of millions of miles. Of a distance so immense the mind can form no adequate conception. The most successful effort towards it is made by gradual and successive approximations. Let us, therefore, take the motion of a rail-way car as the most rapid with which we are familiar, and apply it first to the planetary spaces, and then to the vast interval that separates these nether worlds from the fixed stars. A rail-way car, travelling constantly night and day at the rate of twenty miles per hour, would make 480 miles per day. At this rate, to travel around the earth on a great circle would require about 50 days, and 500 days to reach the moon. If we took our departure from the sun, and journeyed night and day, we should reach Mercury in a little more than 200 years, Venus in nearly 400, and the Earth in 547 years; but to reach Neptune, the outermost planet, would require 16,000 years. Great as appear the dimensions of the solar system, when we imagine ourselves thus borne along from world to world, yet this space is small compared with that which separates us from the fixed stars; for to reach 61 Cygni it would take 324,000,000 years. But this is believed, for certain satisfactory reasons, to be one of the nearest of the stars. Several other stars whose parallax has been determined are at a much greater distance than 61 Cygni. The pole star is five times as far off; and the greater part of the stars are at distances inconceivably more remote. Such, especially, are those which compose the faintest nebulæ. DISCOVERY OF THE PLANET NEPTUNE.—From the earliest ages down to the year 1781, the solar system was supposed to terminate with the planet Saturn, at the distance of nine hundred millions of miles from the sun; but the discovery of Uranus added another world, and doubled the dimensions of the solar system. It seemed improbable that any more planets should exist at a distance still more remote, since such a body could hardly receive any of the vivifying influences of the central luminary. Still, certain irregularities to which the Uranus was subject, led to the suspicion that there exists a planet beyond it, which, by its attractions, caused these irregularities. Impressed with this belief, two young astronomers of great genius, Le Verrier, of France, and Adams, of England, applied themselves to the task of finding the hidden planet. The direction in which the disturbed body was moved afforded some clue to the part of the heavens where the disturbing body lay concealed; the kind of action it excited at different times indicated that it was beyond Uranus, and not this side of that planet; and the magnitude of the forces it exerted gave some intimation of its size and mass. The law of distances from the sun which the superior planets observe (Saturn being nearly twice the distance of Jupiter, and Uranus twice that of Saturn), led both these astronomers to assume that the body sought was nearly double the distance of Uranus from the sun. With these few and imperfect data, as so many leading-strings proceeding from the planet Uranus, they felt their way into the abysses of space by the aid of two sure guides—the law of gravitation and the higher geometry. Both astronomers arrived at nearly the same results, although they wrought independently of each other, and each, indeed, without the knowledge of the other. Le Verrier was the first to make public his conclusions, which he communicated to the French Academy at their sitting, August 31, 1846. They saw that there existed, at nearly double the distance of Uranus from the sun, a planet larger than that body; that it lay near a certain star seen at that season in the southwest, in the evening sky; that, on account of its immense distance, it was invisible to the naked eye, and could be distinctly seen with a perceptible disk only by the most powerful telescopes; being no brighter than a star of the ninth magnitude, and subtending an angle of only three seconds. Le Verrier communicated these results to Dr. Galle, of Berlin, with the request that he would search for the stranger with his powerful telescope, pointing out the exact spot in the heavens where it would be found. On the same evening, Dr. Galle directed his instrument to that part of the heavens, and immediately the planet presented itself to view, within one degree of the very spot assigned to it by Le Verrier. Subsequent investigations have shown that its apparent size is within half a second of that which the same sagacious mind foresaw, and that its diameter is nearly equal to that of Uranus, being 31,000, while Uranus is 35,000 miles.[18] The distance from the sun is less than was predicted, being only about 3000, instead of 3600 millions of miles; and its periodic time is 164½, instead of 217 years, as was supposed by Le Verrier. One satellite only has yet been discovered, and this was first seen by Professor Bond with the great telescope of Harvard University.

RECENT TELESCOPIC DISCOVERIES.—The great reflecting telescope of Lord Rosse, and the powerful refracting telescopes of Pulkova and Cambridge, have opened new fields of discovery to the delighted astronomer. A new satellite has been added to Saturn, first revealed to the Cambridge instrument, making the entire number of moons that adorn the nocturnal sky of that remarkable planet no less than eight. Still more wonderful things have been disclosed among the remotest *Nebulæ*. A number of these objects before placed among the irresolvable nebulæ, and supposed to consist not of stars, but of mere nebulous matter, have been resolved into stars; others, of which we before saw only a part, have revealed themselves under new and strange forms, one resembling an animal with huge branching arms, and hence called the *crab* nebula; another imitating a scroll or vortex, and called the *whirlpool* nebula; and other figures, which to ordinary telescopes appear only as dim specks on the confines of creation, are presented to these wonderful instruments as glorious firmaments of stars.

In the year 1833, Sir John Herschel left England for the Cape of Good Hope, furnished with powerful instruments for observing the stars and nebulæ of the southern hemisphere, which had never been examined in a manner suited to disclose their full glories. This great astronomer and benefactor to science devoted five years of the most assiduous toil in observing and delineating the astronomical objects of that portion of the heavens. He had before extended the catalogue of nebulæ begun by his illustrious father, Sir William Herschel, to the number of 2307; and beginning at that point, he swelled the number, by his labors at the Cape of Good Hope, to 4015. He extended also the list of double stars from 3346 to 5449, and showed that the luminous spots near the South Pole, known to sailors by the name of the "Magellan Clouds," consist of an assemblage of several hundred brilliant nebulæ.

The United States have contributed their full share to the recent progress of astronomy. Powerful telescopes have been imported, made by the first European artists, and numerous others, of scarcely inferior workmanship and power, have been produced by artists of our own. The American astronomers have also been the first to bring the electric telegraph into use in astronomical observations; electric clocks have been so constructed as to beat simultaneously at places distant many hundred miles from each other, and thus to furnish means of determining the difference of longitude between places with an astonishing degree of accuracy; and facilities for recording observations on the stars have been devised which render the work vastly more rapid as well as more accurate than before. Indeed, the inventive genius for which Americans have been distinguished in all the useful arts seems now destined to be equally conspicuous in promoting the researches of science.

FOOTNOTES:

[1] A small pair of globes, that will answer every purpose required by the readers of these Letters, may be had of the publishers of this Work, at a price not exceeding ten dollars; or half that sum for a celestial globe, which will serve alone for studying astronomy.

[2] From two Greek words, τηλε, (*tele,*) *far*, and σκοπεω,(*skopeo,*) *to see.*

[3] Brewster's Life of Newton

[4] Bonnycastle's Astronomy.

[5] Library of Useful Knowledge: History of Astronomy, page 95.

[6] Sir J. Herschel.

[7]A tangent is a straight line touching a circle, as A D, in Fig. 28

[8] Dick's 'Celestial Scenery,' Chapter IV

[9] Dick's 'Celestial Scenery.'

[10]

"As when the moon, refulgent lamp of night, O'er heaven's clear azure sheds her sacred light, When not a breath disturbs the deep serene, And not a cloud o'ercasts the solemn scene, Around her throne the vivid planets roll, And stars unnumbered gild the glowing pole; O'er the dark trees a yellower verdure shed, And tip with silver every mountain's head; Then shine the vales, the rocks in prospect rise, A flood of glory bursts from all the skies; The conscious swains, rejoicing in the sight, Eye the blue vault, and bless the useful light."

Pope's Homer.

[11]The exact longitude of the City Hall, in the city of New York, is 4h. 56m. 33.5s.

[12] You will recollect that the sun is said to be at the node, when the places of the node and the sun are both projected, by a spectator on the earth, upon the same part of the heavens.

[13] Altissimum planetam tergeminum observavi. Or, as transposed, Smaismrmilme poeta leumi bvne nugttaviras.

[14] In imitation of Galileo, Huyghens announced his discovery in this form: a a a a a a a c c c c c d e e e e e g h i i i i i i i l l l l m m n n n n n n n n n o o o o p p q r r s t t t t t u u u u u; which he afterwards recomposed into this sentence: *Annulo cingitur, tenui, plano, nusquam cohærente, ad eclipticam inclinato.*

[15] Dick's 'Celestial Scenery.'

[16] Dick's 'Celestial Scenery.'

[17] The names of all the asteroids known at present are as follows:

1. Ceres.	9. Metis.	17. Psyche.
2. Pallas.	10. Hygeia.	18. Melpomene.
3. Juno.	11. Parthenope.	19. Fortuna.
4. Vesta.	12. Victoria.	20. Massalia.
5. Astræa.	13. Egeria.	21. Lutetia.
6. Hebe.	14. Irene.	22. Calliope.
7. Iris.	15. Eunomia.	23. Un-named.
8. Flora.	16. Thetis.	

[18] Sir John Herschel, however, states its diameter at 41,500 miles

Lightning Source UK Ltd.
Milton Keynes UK
UKHW040835100123
415051UK00021B/431